# Behaviour and Design of Steel and Composite Connections in Fire

Under fire conditions, the strong interactions in structures result in different load-carrying mechanisms and drastic redistributions of internal forces in structural members, which are concentrated at and transferred via connections. Fire safety depends on the performance of these connections, including their temperature distribution and load-carrying mechanisms, and good performance ensures structural robustness in fire.

*Behaviour and Design of Steel and Composite Connections in Fire* is the only dedicated book on fire performance of connections in steel and composite structures. Recent experimental and numerical studies, from individual elements to whole, real-scale structures, have indicated that connections are among the most vulnerable and critical parts of these structures. This book synthesises the research findings on this important subject and explains the essential features in an accessible way in one single source.

The book is ideal for researchers, structural engineers and fire protection engineers in their applications of performance-based fire engineering.

# Behaviour and Design of Steel and Composite Connections in Fire

Mostafa Jafarian
Yong Wang

**CRC Press**
Taylor & Francis Group
Boca Raton London New York

CRC Press is an imprint of the
Taylor & Francis Group, an **informa** business

Cover image: Mostafa Jafarian

First edition published 2023
by CRC Press
4 Park Square, Milton Park, Abingdon, Oxon, OX14 4RN

and by CRC Press
6000 Broken Sound Parkway NW, Suite 300, Boca Raton, FL 33487-2742

© 2023 Mostafa Jafarian and Yong Wang

CRC Press is an imprint of Informa UK Limited

*British Library Cataloguing-in-Publication Data*
A catalogue record for this book is available from the British Library

*Library of Congress Cataloging-in-Publication Data*
Names: Jafarian, Mostafa, author. | Wang, Yong, author.
Title: Behaviour and design of steel and composite connections in fire /
Mostafa Jafarian and Yong Wang.
Description: First edition. | Boca Raton : CRC Press, [2023] |
Includes bibliographical references and index.
Identifiers: LCCN 2022040480 | ISBN 9780367681487 (hbk) |
ISBN 9780367681494 (pbk) | ISBN 9781003134466 (ebk)
Subjects: LCSH: Building, Fireproof. | Steel—Thermal properties. |
Steel, Structural. | Bolted joints—Reliability. | Welded joints—Reliability. |
Fire resistant materials.
Classification: LCC TH1065 .J34 2023 | DDC 693.8/2—dc23/eng/20221109
LC record available at https://lccn.loc.gov/2022040480.

ISBN: 978-0-367-68148-7 (hbk)
ISBN: 978-0-367-68149-4 (pbk)
ISBN: 978-1-003-13446-6 (ebk)

DOI: 10.1201/9781003134466

Typeset in Sabon
by codeMantra

# Contents

# Preface

Since pioneering research of steel and composite structures in fire more than 50 years ago, it is possible to claim that understanding of the behaviour of steel and composite structural frames in fire has reached a good level of maturity. This has enabled the economic gains of undertaking performance-based fire engineering of steel and composite structures to be mostly achieved without compromising structural fire resistance in normal fire limit state design.

In contrast, due to a lack of strong motivation and urgency, research on steel and composite connections in fire lagged behind by decades. There were valid reasons. Connections are usually protected to the highest amount of fire protection of the connected members, and they were not considered weakness of the structure in fire: there are more materials in the connection region than in the connected members, and therefore, temperatures in the connection region are lower than those in the connected members. The implicit assumption is that connections have higher fire resistance than the connected members as long as internal forces in the structure do not change in fire, and the connected members have been designed to possess sufficient fire resistance. Earlier research studies focused on exploring additional connection bending resistance in fire compared to the assumed pin connection, similar to the philosophy of semi-rigid/partial-strength connection design at ambient temperature. However, any gain (e.g. increasing the limiting temperature of the connected beam) from this line of research was rather modest and the interest in further research dwindled.

The Cardington structural fire research program in the middle 1990s in the United Kingdom and fire-induced collapse of the World Trade Center buildings on 11 September 2001 changed the landscape of research on structural behaviour in fire. In a fire situation, there are strong interactions in any structure throughout the fire attack, and these interactions result in drastic changes in internal forces in connections. It is imperative to understand how the internal forces in connections change throughout the entire period of fire attack, including the cooling period, and how they can be dealt with in fire engineering design of structures to avoid structural collapse, in particular, disproportionate collapse, in fire.

The renewed incentive to thoroughly understand connection behaviour in fire has generated a flourish of more recent research activities from about 20 years ago. However, their impact on the practice of performance-based fire engineering of steel and composite structures is still limited. This can be attributed not only to the specialist and complex nature of this topic but also to the wide scatter of information in the literature, principally a range of academic journals. These are barriers to practice. Removing these barriers is the motivation for writing this book: to synthesise research findings on this important subject and to explain its essential features in an accessible way in one single source.

The authors of this book have been active in research and practice of structural fire engineering for many years, many of which are devoted to steel and composite connections in fire. We hope that this book will be an indispensable reference to researchers of this important field, and a valuable guide to fire protection engineers in their applications of performance-based fire engineering to ensure structural safety in fire under different fire conditions.

Yong Wang, Mostafa Jafarian
*July 2022*

# Authors

**Dr Mostafa Jafarian** is a Charted Fire Engineer and works as Technical Director at Hydrock-fire safety division and specialises in structural fire protection and façade behaviour under fire.

He obtained a bachelor's degree in civil engineering, an MSc in Structural Steel Design at Imperial College London, and a PhD in Structural Fire Engineering at the University of Manchester. His PhD project formed part of the RFCS (Research Fund for Coal and Steel) funded European project COMPFIRE investigating the robustness of steel and composite connections in fire.

After obtaining a PhD, he worked as a Research Associate for about two years at the University of Manchester developing a method to specify intumescent coating for concrete filled tubular columns.

Prior to joining Hydrock-fire safety division, he was at Warringtonfire working with different manufacturers and contractors to assist them to evaluate the performance of various structural fire protection systems and façade systems under fire conditions.

He is a member of the British Standards Institution (BSI) committee that coordinates Eurocodes structural fire design and worked with the Association of Specialist Fire Protection (ASFP) (TG1-for reactive fire protection and TG2- for non-reactive fire protection). He is also working closely with BSI committees developing standards related to façade test and structural behaviour under fire conditions.

He also has written and contributed numerous articles to different books and papers covering different aspects of structural and fire engineering.

**Professor Yong Wang** earned a PhD in Structural Engineering from the University of Sheffield in 1988. He is Professor of Structural and Fire Engineering at the University of Manchester. He joined the University of Manchester in 1997 as the leader of structural fire engineering research. Prior to joining the University of Manchester, he spent eight years at the Building Research Establishment, carrying out pioneering research on structural behaviour under fire conditions and was a core member of the BRE team that conducted the Cardington structural fire research programme.

He is a member of a number of British Standards Institution committees responsible for structural Eurocodes for steel/concrete composite structures and fire safety design of structures. He was the UK-nominated expert in three project teams that recently completed the three phases of revision of structural Eurocodes for fire safety design of steel (EN 1993-1-2) and steel/concrete composite (EN 1994-1-2) structures. He currently serves on the legacy 'reference group' for these Eurocodes.

Professor Wang is an internationally renowned researcher on all aspects of structural fire behaviour and has successfully supervised some 50 PhD students, more than 10 post-doctoral researchers, and published more than 400 books, book chapters, design guides and academic papers on the subject.

# List of Symbols

| | |
|---|---|
| $A$ | gross cross-sectional area of beam |
| $A_b$ | nominal area of bolt shank |
| $A_s$ | shank area of bolt |
| $A_m/V$ | section factor of unprotected steel member |
| $A_m$ | surface area of cross-section per unit length |
| $a_{wc}$ | distance between web stiffeners |
| $b_{eff}$ | effective length for column web in compression |
| $c_a$ | specific heat of steel |
| $d$ | distance along the chord of a truss from the support to the brace member of interest |
| $d_b$ | bolt diameter |
| $E$ | Young's modulus of steel plate |
| $E_b$ | Young's modulus of bolt |
| $E_{bt}$ | 1% of the Young's modulus of bolt |
| $E_t$ | 1.5% of the Young's modulus of plate |
| $E_{fi,d}$ | design effect of actions for the fire design situation |
| $F_{truss-centre,0}$ | compression force in the centre brace member of a truss at ambient temperature under the fire limit state design loads |
| $F_{maximum\ chord\ compression}$ | maximum compressive force in the chord member at the centre of a truss at ambient temperature under the fire limit state design loads |
| $F_{truss-centre}$ | total compressive force in the brace member at the centre of a truss for fire design |
| $F_{other\ brace\ member}$ | total compressive force in other brace members of a truss |
| $f_y$ | yield stress of steel |
| $f_u$ | ultimate tensile stress of steel |
| $f_{ub}$ | ultimate tensile stress of bolt |
| $h_{net}$ | net heat flux per unit area |
| $I$ | second moment of area |
| $K$ | stiffness |

| | |
|---|---|
| $K_{eq}$ | stiffness of restraint |
| $K_{br}$ | bearing stiffness |
| $K_b$ | bending stiffness |
| $K_v$ | shearing stiffness |
| $k_{y,\theta}$ | reduction factor for the yield stress of steel at temperature $\theta_a$ |
| $k_{y,\theta}$ | reduction factor for the Young's modulus of steel at temperature $\theta_a$ |
| $k_{sh}$ | correction factor for shadow effect |
| $l_b$ | beam span |
| $L_b$ | effective length of bolt |
| $L_f$ | fin plate length |
| $L_{c1}$ | web cleat length in contact with beam |
| $L_{c2}$ | the web cleat length in contact with column |
| $M_E$ | total external applied moment in beam under pin supports at ends |
| $M_S$ | sagging moment resistance of beam at mid-span |
| $M_h$ | hogging moment resistance of beam at connection |
| $M_{Rd}$ | plastic bending moment resistance of beam at ambient temperature |
| $M_{y, fi, Rd}$ | maximum applied moment in beam |
| $M_{fi, \theta, Rd}$ | plastic bending moment resistance of beam in fire |
| $N$ | axial load in beam |
| $N_{b, fi, t, Rd}$ | compressive resistance of beam in fire |
| $n_{th}$ | number of threads per unit length of bolt |
| $P$ | lateral force |
| $R_{fi, d, 0}$ | design resistance of steel member, for fire design situation, at time $t = 0$ |
| $t_f$ | fin plate thickness |
| $t_w$ | beam web thickness |
| $t_{cf}$ | column flange thickness |
| $t_c$ | web cleat thickness |
| $t_e$ | endplate thickness |
| $V_{Ed, fi}$ | applied shear force in fire |
| $V$ | volume of member per unit length |
| $W$ | section modulus |
| $W_e$ | endplate width |
| $Wc_f$ | column flange width |
| $\alpha$ | coefficient of thermal elongation of steel |
| $\delta$ | deflection of beam |
| $\delta_{peak}$ | deflection of beam at peak catenary force |
| $\Delta y$ | deflection at yield of connection component |
| $\Delta u$ | ultimate deflection of connection component |
| $\Delta\delta_{cl}$ | additional deflection of T-stub after yield |

| | |
|---|---|
| $\Delta\theta$ | average temperature increase in beam |
| $\Delta\theta_a$ | temperature increase of steel section |
| $\Delta\theta_{g,t}$ | increase in fire temperature during time interval $\Delta t$ |
| $\Delta t$ | time interval |
| $\varepsilon_{u,b}$ | ultimate strain of bolt |
| $\varepsilon_u$ | ultimate strain of steel |
| $\theta_{buckling}$ | buckling critical buckling temperature of beam |
| $\theta_{a,cr}$ | critical temperature of beam |
| $\theta_{peak}$ | temperature at peak catenary force |
| $\theta_{trial}$ | trial temperature |
| $\rho_a$ | density of steel |
| $\varphi$ | rotation at connection |

# Chapter 1

# Introduction

## 1.1 INTRODUCTION TO CONNECTION BEHAVIOUR UNDER FIRE

Connections in any structure are the most critical members of the structure. As the saying goes: any fool can design a structure, it takes an engineer to design a connection. A connection transfers forces from one structural member to another, and therefore, understanding structural interactions is crucial. This is particularly challenging for steel and composite framed structures in fire because the forces transmitted through connections when a framed structure is exposed to fire can be drastically different from those at ambient temperature, and these forces are variable throughout the fire exposure.

The complexity of connection behaviour in realistic steel and composite framed structures in fire is best illustrated with reference to the well-publicised full-scale structural fire tests at Cardington in the United Kingdom in the mid-1990s.

Figure 1.1 shows two connections of the Cardington steel-framed structure after fire testing, one with fractured bolts of a fin-plate connection and the other with a severely buckled and distorted lower flange and a web near an end-plate connection. These connections would have been designed and constructed for resisting vertical shear forces only. However, the bolt fracture shown in Figure 1.1a was caused by a horizontal shear force during cooling due to restrained thermal contraction. The distortion of the web and the lower flange in Figure 1.1b was a result of combined action of shear and axial forces. The qualitative reasons why the forces transmitted through connections in fire are different from those at ambient temperature and why they vary are now well understood.

Despite the importance of connections, books on fire safety of steel and composite structures have only scant coverage of this topic (Wang, 2002, Wang et al., 2012, Buchanan and Abu, 2017, Franssen et al., 2009). This reflects a historical lack of adequate knowledge that is only now being gradually improved. It also reflects how the subject of connection behaviour in fire has been researched. As explained in a review paper by Wang (2011),

DOI: 10.1201/9781003134466-1

(a)

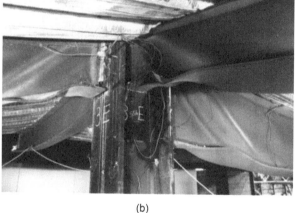

(b)

*Figure 1.1* Failure modes observed during Cardington structural fire tests (Al-Jabri et al., 2008). (a) Bolt fracture during cooling due to thermal contraction. (b) Distortion of web and lower flange due to combine shear and axial load.

connections in steel and composite framed structures are typically considered simple connections and are designed to resist shear forces only. However, as has been established through extensive research studies at ambient temperature, nominally simple connections can offer some bending resistance. Therefore, early research studies have attempted to explore this reserve in bending resistance of connections, focusing on connection behaviour under shear forces and bending moments, without considering variable forces in connections in fire and without including axial forces (Lawson, 1990, Al-Jabri et al., 2008, Leston et al., 1997).

Different behaviours of the Cardington steel-framed building connections and the role connections played in the collapse of the World Trade Center buildings were catalysts for the renewed interest in research on connection behaviour in fire. Since the early 2000s, a lot of research efforts have been devoted to understanding steel and composite connections in fire. Although knowledge on this subject has increased exponentially, the findings of these studies are scattered in academic journals and in a form that is both difficult to access and implement by structural and fire protection engineers. The aim of the book is to provide collective knowledge of this subject in one single publication, this book is intended to make the information accessible to all those who are concerned with structural safety in fire.

## 1.2 INTRODUCTION TO CONNECTION DESIGN FOR FIRE SAFETY

Ensuring structural integrity of connections in fire can only be achieved if the demand on the connection does not exceed resistance of the connection. Therefore, in addition to evaluating the variable forces transmitted through the connection, which will be explained in detail in Chapter 2 of this book, connection resistance should also be accurately assessed. The latter involves two key steps: calculation of temperatures at different connection locations and calculation of connection resistance at elevated temperatures.

The provision in Eurocode 3 Part 1–2 (CEN, 2005a) for evaluating temperatures at different connection locations is very limited, and reference is made to the following three methods:

1. Temperature distribution along the depth of connection as a proportion of the lower flange of the connected beam
2. Simplified calculation method using section factors of connection components
3. Advance methods (using finite element analysis)

Studies by various researchers (e.g. Wald et al., 2009, Ding, 2007, etc.) have shown that the first option is very inaccurate. The second option is simple yet flexible, and many researchers have carried out investigations on how to obtain section factors for a variety of connection components (Dai et al., 2007, 2010, Ding and Wang, 2009, Wald et al., 2009 ). It is important to understand the background, accuracy and applicability of this method. Method 3 requires extensive expertise and is time-consuming to conduct. Therefore, it is more commonly used as a research tool.

Having obtained temperatures of connection components, connection resistance can be evaluated. The current version of Eurocode 3 Part 1–2 (CEN, 2005a) only covers a limited number of design checks for components of connections at elevated temperature, including bolts in shear, bearing and tension, and welds under shear. Whilst such checks are necessary, they are not sufficient for calculating connection resistances under different types of forces (tension, shear, bending) under single action or a combination of actions. As with temperature calculations, advanced finite element methods can be used to evaluate connection resistance. However, usage of FE modelling is not generally a practical approach. The alternative relatively simple, yet flexible approach for practical applications is the component-based method. Briefly, the component-based method is a model whereby a connection is divided into a number of spring components. The behaviour of each spring component is characterised by a force–displacement relationship as a function of temperature. The type of force–displacement relationship of a component differs depending on the type of force in the component. The force–deformation relationships of all the springs are then assembled to characterise the overall connection behaviour under different actions or their combinations (Block et al., 2013, Spyrou et al., 2002, Jafarian and Wang, 2015). To implement this method, it is necessary to obtain connection component temperatures (mentioned in the previous section) and component force–deformation relationships at elevated temperatures.

The component-based model is illustrated in Figure 1.2. For ambient temperature applications, this method is very well developed, and all the necessary information is provided in Eurocode 3 Part 1–8 (CEN, 2005b). However, Eurocode 3 Part 1–8 (CEN, 2005b) for ambient temperature applications only deals with bending moment. For elevated temperature applications, connections can be subjected to combined bending moment and axial force. However, Eurocode 3 Part 1–2 (CEN, 2005a) does not provide guidance about how to use this method. Furthermore, there is no single source of information for obtaining connection component force–displacement relationships at elevated temperatures.

To summarise, there is a need for a single source of information to provide a detailed background understanding of connection behaviour in fire, including structural analysis to obtain variable connection forces in fire, accurate methods of calculating connection component temperatures, reliable material property data, availability of connection component force–displacement relationships and well-constructed examples to demonstrate how to use the component-based method. These are the main topics of this book.

## 1.3 INTRODUCTION TO THE BOOK

This book has seven chapters. Chapter 2 is further divided into two parts. Part 1 presents a detailed summary of realistic connection behaviour in fire, based on a review of existing fire tests and numerical simulations of large-scale structures in fire. Particular foci of this review are interactions

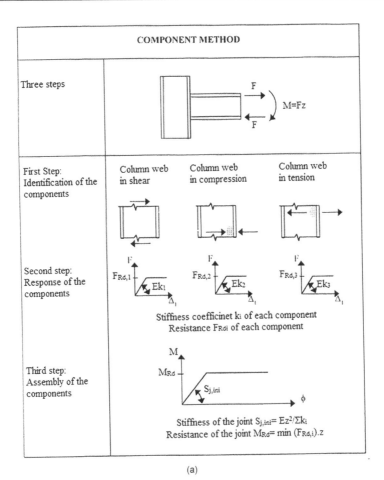

(a)

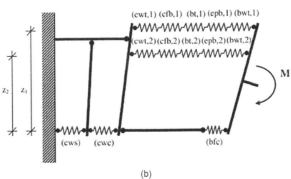

(b)

*Figure 1.2* Illustration of the component-based model for connection design. (a) Three-step approach for using the component based method (Jaspart, 2000). (b) Example of component-based model for an extended end-plate connection (Block, 2006).

between different structural members and hence evolution of variable connection forces and connection failure modes. Part 2 of this chapter explains how the variable connection forces may be obtained in structural analysis.

Chapter 3 evaluates different methods of calculating connection temperatures and explains how to use the simple and yet flexible method (option 2 in Section 1.2) to obtain connection temperatures in different components.

Chapter 4 presents details of the component-based method, including identification of connection components and sources and methods of evaluating force–displacement relations of connection components at elevated temperatures.

Chapter 5 combines the theoretical developments presented in this book and provides a few carefully constructed examples to demonstrate how connection resistance may be checked to resist combined bending moment and catenary action that may develop in the connected beam in fire.

Chapter 6 covers the more special topic of welded steel tubular connections in fire.

Connection behaviour underpins structural behaviour in fire, in particular, ductile connection behaviour is key to ensuring structural robustness under fire conditions. This is a very active research topic, and Chapter 7 will explain the role of connection behaviour in influencing structural robustness in fire and how to improve connection behaviour, not only by increasing their resistance but also their deformation capacity.

## REFERENCES

Al-Jabri, K. S., Davison, J. B. & Burgess, I. W. 2008. Performance of beam-to-column joints in fire - A review. *Fire Safety Journal*, 43, 50–62.

Block, F., Davison, J., Burgess, I. & Plank, R. 2013. Principles of a component-based connection element for the analysis of steel frames in fire. *Engineering Structures*, 49, 1059–1067.

Block, F. M. 2006. *Development of a Component-Based Finite Element for Steel Beam-to-Column Connections at Elevated Temperatures*. University of Sheffield, Sheffield, UK.

Buchanan, A. H. & Abu, A. K. 2017. *Structural Design for Fire Safety*, John Wiley & Sons, Chichester.

CEN 2005a. EN 1993-1-2 Eurocode 3: Design of Steel Structures–Part 1–2: General Rules–Structural Fire Design. British Standards Institution.

CEN 2005b. EN 1993-1-8:2005. Part 1–8, Design of Joints. Brusels: British Standards Institution.

Dai, X., Wang, Y. & Bailey, C. 2007. Temperature distribution in unprotected steel connections in fire. *Steel and composite structures*, Taylor and Francis Group, London.

Dai, X., Wang, Y. & Bailey, C. 2010. A simple method to predict temperatures in steel joints with partial intumescent coating fire protection. *Fire Technology*, 46, 19.

Ding, J. 2007. *Behaviour of Restraint Concrete Filled Tubular (CFT) columns and their joints in fire*. PhD, Manchester.

Ding, J. & Wang, Y. C. 2009. Temperatures in unprotected joints between steel beams and concrete-filled tubular columns in fire. *Fire Safety Journal*, 44, 16–32.

Franssen, J. M., Kodur, V. & Zaharia, R. 2009. *Designing steel structures for fire safety*, CRC Press, Boca Raton, FL.

Jafarian, M. & Wang, Y. C. 2015. Force–deflection relationship of reverse channel connection web component subjected to transverse load. *Journal of Constructional Steel Research*, 104, 206–226.

Jaspart, J. P. 2000. General report: Session on connections. *Journal of Constructional Steel Research*, 55, 69–89.

Lawson, R. M. 1990. Behaviour of steel beam to column connections in fire. *Structural Engineering*, 68, 8.

Leston, C, Burgess, I., Lennon, T. & Plank, R. 1997. Elevated-temperature moment-rotation tests on steelwork connections. *Proceedings of the Institution of Civil Engineers-Structures Buildings*, 122, 410–419.

Spyrou, S., Davison, J., Burgess, I. & Plank, R. 2002. Component studies for steelwork connections in fire. *5th International Conference on Stability and Ductility of Steel Structures*, Budapest, Hungary, 769–776.

Wald, F., Sokol, Z. & Moore, D. 2009. Horizontal forces in steel structures tested in fire. *Journal of Constructional Steel Research*, 65, 1896–1903.

Wang, Y., Burgess, I., Wald, F. & Gillie, M. 2012. *Performance-Based Fire Engineering of Structures*, CRC Press, Boca Raton, FL.

Wang, Y. C. 2002. *Steel and Composite Structures*. First ed. Taylor & Francis, London and New York.

Wang, Y. C. 2011. Performance based fire engineering research of steel and composite structures: A review of joint behaviour. *Advances in Structural Engineering*, 14, 613–624.

# Chapter 2

# Structural analysis of connections in fire

The aim of safe design of structures is to ensure that every part of the structure has sufficient resistance to the internal forces developed in the structure as a result of external actions on the structure. This applies to the design of connections in fire. As will be demonstrated in the following sub-section, different types of internal force (axial force, shear force, bending moment) co-exist in the connection, and these internal forces vary over time and follow complex patterns. Therefore, it is important to identify the key stages of connection behaviour in fire so as to be able to accurately quantify the internal forces in connections at these key stages. This is the focus of this chapter. In Chapter 5, methods will be presented for checking connection resistance to resist such different internal forces.

## 2.1 A BRIEF SUMMARY OF CONNECTION BEHAVIOUR IN FIRE

Early research studies on connection behaviour in fire focused on observing connection failure modes and quantifying connection resistance at elevated temperatures under **defined** internal forces in the connection. Prominent research studies of this type include those of Lawson (1990), Leston Jones et al. (1997) and Al-Jabri (1999), which investigated cruciform connection arrangement, as shown in Figure 2.1. In this type of arrangement, the connection structure is statically determinate. The shear force and bending moment in the connection are unchanged throughout the entire heating duration, and there is no axial force in the connection. The context of these research studies was similar to that at ambient temperature: to quantify the bending resistance of the connection in fire to explore how the connection may benefit fire resistance design of the connected beams, following the tradition of semi-rigid connection research at ambient temperature in the preceding 30 years or so (Nethercot, 2006).

While these research studies could not help with quantifying the varying internal force conditions in connections, they were helpful in understanding different failure modes of connections (e.g. Figure 2.1) and in developing

DOI: 10.1201/9781003134466-2

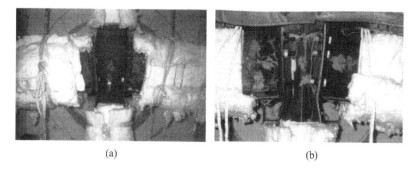

(a)                                                    (b)

*Figure 2.1* Cruciform arrangement of connection and typical failure modes under bending moment (Leston-Jones et al., 1997). (a) Column web failure. (b) Column web and flange failure.

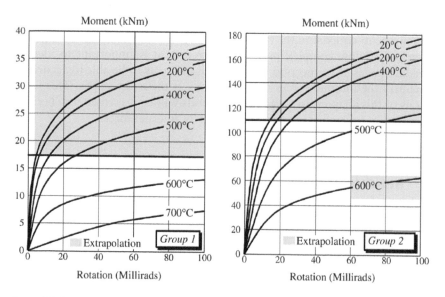

*Figure 2.2* Moment-rotation curves of flush endplate connections at elevated temperature (Al-Jabri et al., 2005).

methods to calculate the resistance of connections to different types of internal forces, in particular, the moment-rotation curves at elevated temperatures, as shown in Figure 2.2.

Two major events profoundly influenced the development of structural fire engineering: the controlled structural fire testing programme at Cardington in the United Kingdom in the mid-1990s and the World Trade Center disaster on 11 September 2001. One key observation of these events was the role of connections: how they critically dictated progressive collapse of structures in fire, how they affected the strong interactions between different structural members and how the internal forces in connections in fire were drastically different from those at ambient temperature.

Various subsequent research studies of connections in fire, in the more realistic context of connections interacting with other structural members, either experimentally, such as Liu et al. (2002), Ding and Wang (2007), Wang et al. (2011) and Lopes et al. (2011), or numerically, such as Yin and Wang (2004) and Ramli-Sulong (2007), have all revealed similar patterns of how the internal forces of connections vary with increasing temperature.

Qualitatively, the full range of connection behaviour in fire and the patterns of variation of connection internal forces at elevated temperatures are, as described by Wang (2002), shown in Figure 2.3. Of course, any connection may fail before going through all the stages of full connection behaviour. The following summarises the complete behaviour of moment-resisting connections under axial and rotational restraints by the surrounding structure:

- When the connected beam is exposed to fire, it will expand and rotate because of high temperature and thermal curvature, respectively. Due to rotational and axial restraints provided by the surrounding columns and connections, free thermal expansion and rotation of the beam are not possible. As a result, the restrained movement and rotation will, respectively, be converted into a compression force $(P)$ and hogging moments $(M_h)$.
- The hogging moment and compression force in the beam will increase with increasing temperature until local (e.g. bearing failure of the web) or temporary global failure of the beam (stage 1).
- Following the temporary failure of the beam, the compression force and hogging moments in the beam start to decrease while the vertical deflection $(\delta_v)$ of the beam accelerates.
- This process continues until the beam's shortening caused by its large vertical deflection compensates its thermal elongation. At the transition stage (stage 2), the compression force in the beam returns to zero, and the hogging moment returns to the original value at the start of the fire.
- After stage 2, the beam enters into catenary action. During this phase of beam behaviour, shortening of the beam due to its large vertical deflection (i.e. $\delta_v$) overtakes its thermal elongation. Because of axial restraint, a tensile force develops in the beam. At the same time, both the hogging and sagging moments of the beam decrease to a negligible value due to an increase in temperature combined with an increase in catenary force.
- The catenary tensile force in the beam continues to increase until reaching the tensile resistance of the beam at elevated temperatures, after which the tensile force in the beam follows the tensile resistance of the beam. During this phase of beam behaviour, the internal moments in the beam are almost 0.

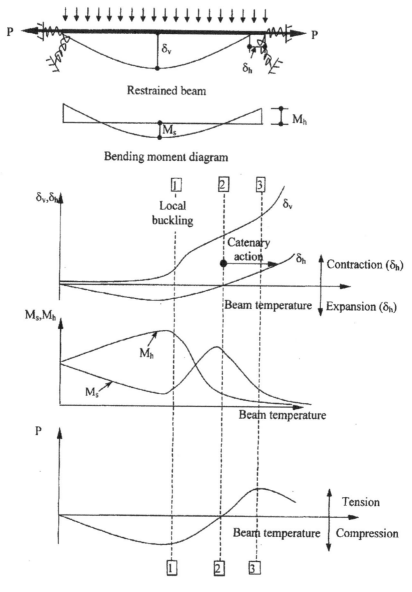

*Figure 2.3* Variations of internal forces and deformations of an axially and rotationally restrained beam during fire (Wang, 2002).

- At any stage of the catenary action phase of beam behaviour, the connection could fail, due to the connection not possessing sufficient rotational capacity to meet the demand of a very large deflection of the beam.

- Throughout all the above stages, the shear force in the connection remains unchanged.
- Upon cooling, the internal forces in the beam, and hence in the connection, follow a new course, as will be described in more detail in Section 2.3.4.

## 2.2 KEY STAGES OF CONNECTION BEHAVIOUR

The internal forces in connections in fire experience complex variations, as summarised above. It would not be possible to accurately trace connection behaviour continuously throughout the entire duration of heating and cooling. Therefore, this book, and this chapter for structural analysis of connections in fire, will only consider a few key stages of connection behaviour and will present simplified methods for calculating connection forces at these key stages. These different key stages are illustrated in Figure 2.4, and they indicate different possible limit states of the connection.

Stage 1: Maximum compression force in the connection

At this stage, the compression force in the connection increases due to restrained thermal expansion until reaching the maximum value and then decreases afterwards due to increased flexibility of the structure at an elevated temperature until returning to zero. Although the compression force may have implications on the connected columns, connection safety in fire is not an issue if no fracture is involved, as is the case with using endplate connections. The only situation where the connection may be of concern due to the compression force is when checking the shear resistance of some non-moment resisting connections, such as fin plate connection, under combined compression force and shear force. In such connections, any connection bending moment can be considered negligible.

Furthermore, for moment-attracting connections such as flush/extended endplate connections or welded connections, there is no need to check any increase in tension force in any connection components as a result of the increase in connection hogging moment. The increases in connection hogging moment and compression force follow a similar trend because they are a result of the same source: bending stiffness of the surrounding structure. Therefore, it can be considered that the increase in tension in any connection component due to increased connection hogging moment is offset by the corresponding increase in compression in the same component due to increased compression force in the connection.

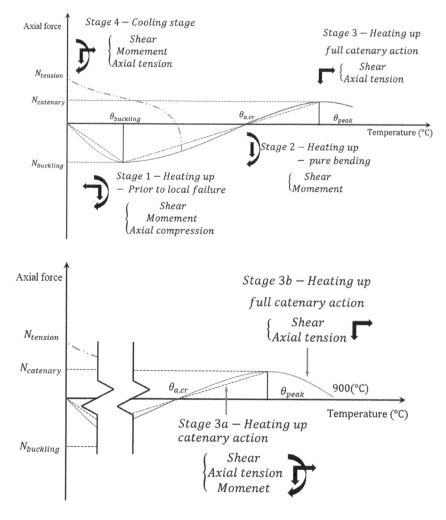

Figure 2.4 Key stages and internal forces of connection to be quantified.

Therefore, for this stage, only the maximum compression force will be calculated. When checking connection safety, this maximum compression force will be combined with the applied shear force for non-moment-resisting connections, such as fin plate connections.

Details of how to calculate this maximum compression force are presented in Subsection 2.3.1.

Stage 2: Bending resistance

This stage marks the transition of connection behaviour from internal compression to catenary action. At this stage, the beam shortening caused by large vertical deflection cancels out the beam thermal expansion so that the internal compression force in the connection is zero, and the bending moments are similar to those at ambient temperature.

This is the conventional stage for which existing fire resistance methods are available to check the bending resistance of the connected beam. No further structural analysis is necessary for the connection.

Stage 3: Maximum catenary force in the connection

During the catenary action stage of the beam, the catenary action force increases with temperature until it reaches the axial tensile resistance of the cross-section of the beam. When this happens, the bending moments in the beam have diminished to a negligible level and the beam may be considered to be in pure catenary action. However, the beam should be able to successfully transit from the pure bending stage to the pure catenary action stage during which there is co-existence of bending moment (if the connection is moment resisting), shear force and axial tension in the connection.

The catenary action stage can be important in helping the structure resist disproportionate collapse. However, it should be appreciated that during the entire stage of catenary action, the structure is operating at the boundary of its resistance to force. How long the structure can survive will depend on whether the connection can enable the connected beam to develop the necessary catenary action when it has reached its rotation capacity. To check connection resistance, it is necessary to calculate the rotation demand on connections. This is presented in Subsection 2.3.3.

Stage 4: Cooling to ambient temperature

A connection can fail during the cooling stage when large tensile forces are generated, as shown in Figure 2.4. The later cooling starts, the greater the residual tensile force in the connection. However, it can be argued that the connected beam is designed, and therefore intended to be employed, to only reach the limit of pure bending (stage 2). Therefore, if it is necessary to consider the behaviour of the connected beam during the catenary action stage (stage 3 above) after reaching the pure bending stage, it is for the purpose of controlling disproportionate collapse. Connection failure during the cooling stage is usually local, concentrated in the component under the highest load. After local failure, the residual tension force in the connection is greatly released. Such local connection failure, when happening when the global structure is recovering its strength during the cooling phase, has a much lower consequence than disproportionate collapse of the global structure during the heating phase. Therefore, it is not necessary to explicitly

Table 2.1 Key internal connection forces to be quantified

| Stage | Connection internal force | Necessary design checks |
|---|---|---|
| Stage 1 (restrained thermal expansion) | Maximum compression force | Shear resistance |
| Stage 2 (pure bending) | Bending moment | Moment capacity of moment resisting connection |
| Stage 3 (catenary action) | Tension and bending moment (if moment resisting connection) | Resistance to tension and bending at full rotation capacity |
| Stage 4 (cooling) | Residual tensile force at ambient temperature, cooling starting from beam pure bending | Tensile resistance at ambient temperature |

check the effect of cooling on connection behaviour during the catenary action stage. Therefore, the critical situation is when connection cooling starts from the pure bending stage of the connected beam. Subsection 2.3.4 will describe how the maximum residual axial force can be calculated after cooling from stage 2.

In summary, Table 2.1 lists the key internal connection forces and their corresponding capacity checks, in addition to shear force, that should be quantified at different stages.

## 2.3 SIMPLIFIED METHODS TO CALCULATE INTERNAL FORCES OF DIFFERENT KEY STAGES OF CONNECTION BEHAVIOUR AT ELEVATED TEMPERATURES

### 2.3.1 Key stage 1: maximum compression force

The maximum compression force in the connection can be calculated with reference to Figure 2.5, which shows an axially restrained member with restrained thermal expansion. The increase in compression force in the beam due to restrained thermal expansion can be calculated as follows:

$$P = K_{eq} \cdot l_b \alpha \Delta\theta \qquad (2.1)$$

where $K_{eq}$ is the stiffness of the restraint (i.e. surrounding columns) to a lateral force $P$ on the assumption that the axial stiffness of the beam is infinite compared to the lateral stiffness of the restraint, $l_b$ is the beam span, $\alpha$ is the coefficient of thermal elongation of steel and $\Delta\theta$ is the average temperature increase in the beam. The maximum compression force is reached

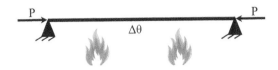

*Figure 2.5* Increase in compression force due to restrained thermal expansion.

when the beam buckles under combined bending and compression, which can be approximated as follows:

$$\frac{N_{fi,\,Ed}}{N_{b,\,fi,\,t,\,Rd}} + \frac{M_{y,\,fi,\,Rd}}{M_{fi,\,\theta,\,Rd}} = 1 \tag{2.2}$$

where

$$N_{fi,\,Ed} = P_{max}$$

$M_{y,\,fi,\,Rd}$ is the maximum applied moment in the beam, which can be assumed to be the same as at ambient temperature using the load combination for fire design

$N_{b,\,fi,\,t,\,Rd} = \chi_{fi} A k_{y,\,\theta} f_y$ is the compressive resistance of the beam under fire, and

$M_{fi,\,\theta,\,Rd} = k_{y,\,\theta} M_{Rd}$ is the plastic bending moment resistance of the beam under fire.

in which

$A$ is the gross cross-sectional area of the beam,

$f_y$ is the yield stress of steel,

$k_{y,\,\theta}$ is the reduction factor for the yield stress of steel at temperature $\theta_a$,

$M_{Rd}$ is the plastic bending moment resistance of the beam at ambient temperature.

Combining equations 2.1 and 2.2 gives

$$K_{eq}.l_b \alpha \Delta \theta = N_{b,\,fi,\,t,\,Rd} \left(1 - \frac{M_{y,\,fi,\,Rd}}{M_{fi,\,\theta,\,Rd}}\right) \tag{2.3}$$

Solving equation 2.3 gives the critical buckling temperature of the beam $(\theta_{buckling})$. Afterwards, equation 2.1 is used to obtain the maximum compression force $P_{max}$.

It is possible that the maximum compression force calculated above greatly exceeds the shear resistance of the connection. If this is the case, the critical failure mode of the connection should be controlled so that after connection failure, the compression force is released but the connection still maintains its integrity. This means that the connection should be designed so that ductile yielding of the connected beam component is the failure mode, not brittle fracture of connection components such as bolt and weld.

### 2.3.2 Stage 2: connection bending moment

At this stage, the connection bending moment (if moment resisting) is assumed to be the same as that at ambient temperature. To check whether the connection has sufficient resistance, it is necessary to find the connection temperature, which is the connection temperature attained at the same time when the connected beam has reached its critical temperature. The beam critical temperature is calculated below according to EN 1993-1-2:

$$\theta_{a,\,cr} = 39.9 \ln \left[ \frac{1}{0.9674 \; \mu_0^{3.833}} - 1 \right] + 482 \tag{2.4}$$

where $\mu_0$ is utilisation factor at $t=0$ and can be calculated as

$$\mu_0 = \frac{E_{fi,\,d}}{R_{fi,\,d,\,0}} \tag{2.5}$$

in which $E_{fi,\,d}$ is the design effect of actions for the fire design situation, according to EN 1991-1-2, and $R_{fi,\,d,\,0}$ is the corresponding design resistance of the steel member, for the fire design situation, at time $t=0$.

Deflection due to beam elongation at this stage can be calculated as Usmani et al. (2001), as follows:

$$\delta = \frac{2l}{\pi} \sqrt{\alpha \Delta \theta + \frac{\alpha \Delta \theta^2}{2}} \tag{2.6}$$

### 2.3.3 Stage 3: bending moment and tensile force

For a connection to be able to reach this stage, it is required to demonstrate a high rotation capacity. However, to evaluate the demand on rotation, it is necessary to know the maximum deflection of the beam. This can be done according to the following equilibrium equation of the beam:

$$N\delta = M_E - \left( M_s + M_b \right) \tag{2.7}$$

where

N is the axial load in the beam (tension positive),

$\delta$ is the maximum total deflection of the beam (total of thermal bowing deflection $\delta_1$ and that induced by mechanical loading $\delta_2$),

$M_E$ is the total external applied moment in the beam under pin supports at ends,

$M_s$ is the sagging moment resistance of the beam at mid-span,

$M_h$ is the hogging moment resistance of the beam at connection.

As explained in Section 2.2, it is assumed that after reaching the maximum value, with increasing temperature, the tension force in the beam follows the tension resistance of the cross-section of the beam, and the sagging and hogging moment resistances of the beam are zero. Therefore, in Figure 2.4, the catenary stage is divided into two sub-stages, before (stage 3a) and after (stage 3b) reaching the peak tensile force.

### 2.3.3.1  Stage 3a (before reaching the peak catenary tensile force)

Within this stage, the axial force increases from zero to the maximum value, while both the hogging (if moment resisting connection) and sagging moments gradually decrease to zero at the temperature of peak tensile force.

To calculate the temperature of peak catenary force, it can be assumed that the load–temperature curve changes in a parabolic function. The parabolic function is determined by the following three points: zero axial force at the critical temperature $\theta_{a,\,cr}$ and at 900°C and the tension resistance of the beam's cross-section at the average temperature of

$$\theta_{\text{peak}} = \frac{\theta_{a,\,cr} + 900}{2} \tag{2.8}$$

According to the yield strength–temperature relation of steel in EN 1993-1-2, the yield strength of steel is 0 at a temperature of 1200°C. However, the yield strength of steel decreases sharply to a very low value (0.06) at 900°C and thereafter reaches an almost plateau of very low value with further increase in temperature.

To check the accuracy of the above method, Table 2.2 compares the calculation results using this simple method with the numerical simulation results of Yin (2004) for a steel section size UB457 × 152 × 60. The above simple method can be considered to give a good estimate of the maximum tensile force and the temperature at which this force is reached.

Table 2.2 Comparison between calculation and simulation results of Yin (2004) for temperature at peak tensile force

| Span (m) | Axial restraint ratio | Utilisation factor ($\mu_0$) | Critical temp- $\theta_{a,cr}$ (°C) | Peak axial force temp. $\theta_{peak\text{-}Yin}$ (°C) | Peak axial force temp. $\theta_{peak\text{-}predict\text{-}equation\ 2.7}$ (°C) | $\theta_{peak\text{-}Yin}/\theta_{peak\text{-}predict}$ -equation 2.7 |
|---|---|---|---|---|---|---|
| 8 | ∞ | 0.4 | 649.5 | 783.0 | 774.8 | 0.989 |
| 6.5 | ∞ | 0.4 | 652.3 | 776.7 | 776.2 | 0.999 |
| 5 | ∞ | 0.4 | 647.5 | 742.4 | 773.7 | 1.04 |
| 8 | ∞ | 0.7 | 545.3 | 703.5 | 722.6 | 1.027 |
| 6.5 | ∞ | 0.7 | 548.1 | 687.0 | 724.1 | 1.054 |
| 5 | ∞ | 0.7 | 538.8 | 659.7 | 719.4 | 1.09 |
| 8 | 0.15 | 0.4 | 656.1 | 798.4 | 778.1 | 0.97 |
| 8 | 0.3 | 0.4 | 655.5 | 808.6 | 777.8 | 0.96 |
| 8 | 1 | 0.4 | 655.0 | 800.0 | 777.5 | 0.97 |
| 8 | 0.15 | 0.7 | 551.0 | 729.2 | 725.5 | 0.99 |
| 8 | 0.3 | 0.7 | 550.0 | 721.8 | 725.0 | 1.00 |
| 8 | 1 | 0.7 | 550.0 | 710.8 | 724.9 | 1.02 |

The beam hogging bending moment at the connection (if moment resisting) can be calculated by linear interpolation, as follows:

$$M_b = \frac{-M_{b,\ cr}}{\theta_{peak} - \theta_{a,\ cr}}\left(\theta - \theta_{a,\ cr}\right) + M_{b,\ cr} \tag{2.9a}$$

$$N = \frac{N_c}{\theta_{peak} - \theta_{a,\ cr}}\left(\theta - \theta_{a,\ cr}\right) \tag{2.9b}$$

where
$N_c$ is the tensile resistance of the cross-section of the beam at $\theta_{peak}$ and $M_{b,\ cr}$ is the beam hogging moment at ends at the critical temperature $\theta_{a,\ cr}$.

### 2.3.3.2 Stage 3b (after reaching peak tensile force)

During this stage, the tensile force in the beam follows the tensile resistance of the cross-section, and the bending moments are zero.

### 2.3.3.3 Demand on rotation

As explained in Section 2.2, the temperature at which the connection fails is determined by whether the connection can enable the beam to develop the required catenary action when it has reached its rotation capacity. This is determined by checking the demand on rotation, which can be calculated as follows.

Refer to the beam equilibrium equation (2.7), the deflection of the beam at the onset of full catenary action (peak tensile force) is

$$\delta = \frac{M_E}{N_c} \tag{2.10}$$

The deflection of the beam prior to reaching the full catenary action stage can be assumed to be linear between the peak compression force and the peak catenary force. Before reaching the peak compression force, the beam deflection may be assumed to be the same as that at ambient temperature or neglected due to its small value.

Assuming a parabolic deflection profile for the beam, the maximum rotation of the beam at the connection (hence, the demand on rotation for the connection) can be calculated as:

$$\varphi = \tan^{-1}\left(\frac{4\delta}{L}\right) \tag{2.11}$$

Based on the key points determined in this section, the complete beam axial force–temperature relationship can be obtained. Two examples are provided to show how the method is implemented. Figure 2.6 compares calculation results with the simulation results of Yin (2004) for beam axial force variation for a number of cases.

Figure 2.7 presents comparison results maximum deflection of the beam.

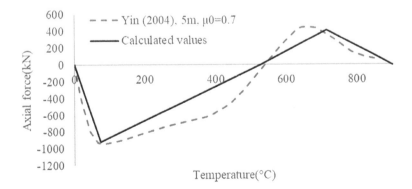

*Figure 2.6* Comparison between calculated results with simulation results of Yin (2004) for axial force.

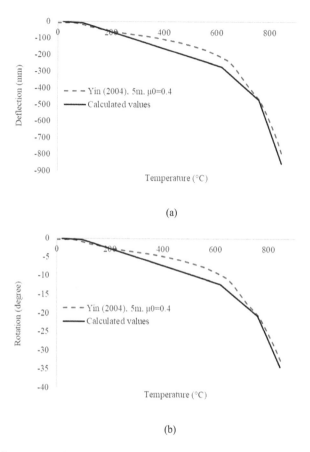

*Figure 2.7* Comparison between calculated deflection and rotation results with simulation results of Yin (2004) for deflection. (a) Comparison between deflections. (b) Comparison between rotations.

### 2.3.4 Stage 4: maximum force at ambient temperature after cooling

According to Ding and Wang (2007), who experimentally and numerically investigated the cooling behaviour of restrained beams in fire, the increase in tensile force in the beam during cooling due to restrained thermal contraction can be considered to follow a parallel path to that of increasing compression force with increasing temperature due to restrained thermal expansion according to equation 2.1 (Figure 2.8). As under the peak compression force, the maximum tension force in the beam may exceed the shear resistance of the connection. If this is the case, the ductile failure mode by yielding of the connected beam, instead of brittle failure mode by bolt fracture or weld failure, should be ensured.

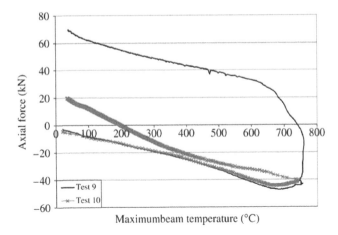

*Figure 2.8* Variation of residual tensile force during cooling from the beam critical temperature for two tests of Ding and Wang (2007).

## 2.3.5 Worked examples

Example 2.1 – Comparison against the simulation results of Yin and Wang (2004) for a rigid connection with infinite axial stiffness

General information
Span: 8 m
Cross-section: UB 457×152×60
A 7620 mm²
I 25500 cm⁴
Applied load in fire:31.04 kN/m

Material properties of steel at ambient temperature
Yield stress $f_y$=275 N/mm²
Young's modulus $E$=205000 N/mm²

Cross-section properties
$A_{eff}$ 6955.7 mm²
$I_{eff, y-y}$ 24341.2 cm⁴
$W_{y-y}$ 1290 cm³
Internal forces in fire
$V_{Ed, fi}$=31.04×8/2=124 kN
$M_{Ed, fi}$ @connection =31.04×8²/12=165.55 kN·m
$M_{Ed, fi}$ @mid span =31.04×8²/24=82.77 kN·m
$M_E$ pin boundary condition=248.32 kN·m
Bending capacity of the beam=275×1290×10³/10⁶ =354.75 kN·m;

Buckling resistance of the beam
For the fixed boundary condition and according to EN1993-1-1 and EN1993-1-2
$N_{b, fi, t, Rd}$=1601.62 kN
$M_{Ed, fi}/M_{Rd, fi,20}$=165.55/345.75=0.467

Resistance to axial compression in the presence of bending: $N_{fi, Ed}=1$
601.62×(1−0.467)=854.21 kN

## Calculation of compression buckling temperature of the beam $\theta_{buckling}$ (Stage 1) – (equation 2.3)

$\alpha=1.2\times10^{-5}$

According to equation 2.3:

$\Delta\theta=N_{fi, Ed}/\ EA_{eff}\ \alpha = 854.21\times103/\ (205000\times6955.7\times1.2\times10^{-5})$

$\Delta\theta=49.92°C$

$\theta_{buckling}=49.92+20=69.92°C$

## Calculation of $M_{Ed}$ the critical temperature $\theta_{a, cr}$ (Stage 2) – (equation 2.4)

$\mu_0=M_{Ed, fi}/M_{Rd,20}=248.32/(2\times354.75)=0.35$

According to equation 2.4:

$\theta_{a, cr}=39.9\ \ln\ [(1/(0.9674\times0.7^{3.833}))-1]+482=640.3°C$

## Calculation of the peak temperature $\theta_{peak}$ and catenary resistance (Stage 3) – (equation 2.7)

From equation 2.7:

$\theta_{peak}=(640.3+900)/2=770.2°C$

$k_{y, \theta}=0.21$

$N_{catenary}=(0.15\times275\times6955.7)/1000=278.9\ kN$

Equation 2.9 gives

$\delta_{catenary}=248.32\times1000/\ 278.9=890\,mm$

### Axial load–temperature graph

| Temp. (°C) | Axial force (kN) | Deflection | Deflection (mm) |
|---|---|---|---|
| 0 | 0 | $WL^4/384E_{20}I$ | $-31.04\times8000^4/205000\times25500\times10^4\times384$ $=-6.33$ |
| 69.92 | −854.21 | $(1/384)WL^4/E_{69.92}\ I$ | $31.04\times8000^4/205000\times25500\times10^4\times384$ $=-6.33$ |
| 640.3 | 0 | $(2L/\pi)\ \sqrt{\varepsilon_T}+0.5\ \varepsilon_T^2$ | $(2\times8000/\pi)\times0.007444+0.5\times0.007444^2$ $=-440.22$ |
| 770.2 | 278.91 | $M/Nc$ | $-248.32\times1000/278.91=-890$ |
| 760 | 256.32 | $M/F_T$ | $-248.32\times1000/256.32=-968.8$ |
| 900 | 0 | - | $\infty$ |

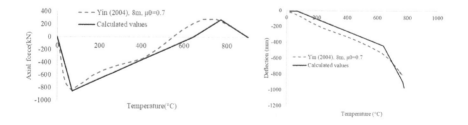

Example 2.2 – An extended endplate connection based on the details in SCI P398 (Brown et al., 2013)

## General information
Span: 6 m
Cross-section: UB 533×210×92
Cross-section: UB 254×254×107

## Applied load
Point load of 324.5 kN at mid-span
$M_{Ed, fi}$ @connection = 324.5×6/8=243.38 kN·m
$M_{Ed, fi}$ @mid-span = 163.62×6/8=243.38 kN·m
$M_E$ pin boundary condition under the applied load in fire 163.62×6/4= 486.75 kN·m

## Material properties of the sections
Yield stress $f_y$=275 N/mm$^2$
Young's modulus $E$=205000 N/mm$^2$

## Cross-section properties under compression
$A_{eff}$=10880.4 mm$^2$
$I_{eff, y-y}$=52794.7 cm$^4$
$W_{y-y}$=2360 cm$^4$

## Buckling resistance of the beam section
For the clamped boundary condition and in line with EN1993-1-1 and EN1993-1-2 (more detailed references can be found in the appendix)
$N_{b, fi, t, Rd}$=2675.53 kN
$M_{Rd,20}$=275×2360×10$^3$/10$^6$=649kN·m
$M_{Ed, fi}/M_{Rd,20}$= 243.38/649=0.38
Resistance to axial compression in the presence of bending: $N_{fi, Ed}$= 2675.53×(1−0.38)=1672.20 kN

## Stiffness of the beam and axial restraint
Axial stiffness of the beam $K_b$=$E$ $A_{eff}$ /$l$=205000×10880.4 / 6000 = 371747.2 kN/m
Axial restraint stiffness: $K_c$=127555.6 kN/m
Equivalent stiffness: $K_{eq}$= $K_b$ $K_c$/($K_b$+$K_c$)=94969.27 kN/m

## Calculation of the buckling temperature $\theta_{buckling}$ (Stage 1) – (equation 2.3)
$\alpha$=1.2×10$^{-5}$
$\Delta\theta$=1672.20/(94969.27×6×1.2×10$^{-5}$)=244.6°C
$\theta_{buckling}$=244.6+20= 264.6°C

## Calculation of the critical temperature $\theta_{a, cr}$ (Stage 2) – (equation 2.4)
$\mu_0$=$M_{Ed, fi}/M_{Rd,20}$=486.75/(1.5×649)=0.5
*Note: this is based on the connection resistance calculations at ambient temperature: the connection can develop a hogging moment that is 50% of the sagging moment in the beam span.
$\theta_{a, cr}$=39.9 ln [(1/(0.9674×0.5$^{3.833}$))−1]+482 =584.7°C

Calculation of the peak temperature $\theta_{peak}$ and catenary resistance (Stage 3) – (equation 2.7)

$\theta_{peak}=(584.7+900)/2=742.3°C$

$k_{y,\theta}=0.1792$

$N_{catenary}=(0.1792\times275\times10880.4)/1000=536.189$ kN

Equation 2.9 gives

$\delta_{catenary}=486.75\times1000/536.19=907.8$ mm

### Axial load–temperature graph

| Temp. (°C) | Axial force (kN) | Deflection | Deflection (mm) |
|---|---|---|---|
| 20 | 0 | $Pl^3/192E_{20}I$ | $-324.5\times6\times10^6/192\times205000\times55200\times10^4=-3.23$ |
| 264.6 | −1672.20 | $Pl^3/192E_{264.6}I$ | $-324.5\times6\times10^6/192\times205000\times0.835\times55200\times10^4=-3.86$ |
| 584.7 | 0 | $(2l/\pi)\times0+0.5\,\varepsilon_T^2$ | $(2\times6000/\pi)\times0.0068+0.5\times0.0068^2=-314.96$ |
| 742.3 | 536.19 | $M/Nc$ | $-486.75\times1000/536.189=-907.8$ |
| 900 | 0 | - | $\infty$ |

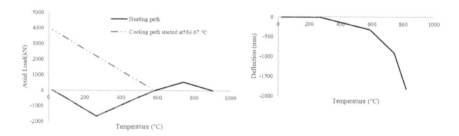

### Cooling stage (Stage 4)

It is assumed that the system underwent the cooling stage when the beam temperature was 584.7°C.

$\theta_{start-cooling}$ 584.7°C

$N_{tension,\,cooling}=(-1672.23/(264.6-20))\times(20-584.7)=3861.06$ kN

## 2.4 NUMERICAL MODELLING

### 2.4.1 Introduction

This chapter has so far described the potentially very complex behaviour of connections in fire and the need to maintain connection integrity at different stages in fire, which entails comparison of actions in the connection (internal force or rotation) against their limits. The main focus of this chapter is developing a simplified calculation method to calculate the key quantities of connection actions (internal forces or rotation demand) so that they can be checked against their respective limit values to demonstrate connection adequacy.

However, many assumptions have necessarily been made. While the simplified calculation method of this chapter should be applicable to many common connections, numerical modelling may be necessary in more specialised cases. Detailed modelling of structural and connection behaviour and failure in fire is challenging and requires a thorough understanding of the fundamentals of structural behaviour in fire and extensive experience in numerical modelling. Providing detailed guidance on such modelling is beyond the scope of this book. Instead, this section will only identify a few of the more challenging issues and provide some qualitative advice. Specifically, these challenges are related to numerical convergence and excessive computation time.

## 2.4.2 Techniques to overcome numerical convergence problems

Non-convergence is one of the most common FE modelling issues. Numerical non-convergence problems can occur under many situations, including material failure (fracture), interface complexity and temporary loss of stability of the structure.

### 2.4.2.1 Material failure (fracture)

Modelling fracture is challenging in the context of simulating large-scale structural behaviour in fire. Therefore, it is advised not to explicitly model material fracture. A possible alternative is to assume that the fractured material retains a very small amount of its resistance and stiffness. This may still cause temporary numerical instability which can be dealt with following the advice in section 2.4.2.2.

### 2.4.2.2 Dealing with temporary loss of stability

One method that could be used to deal with numerical instability is employing the displacement control solver. This may allow the numerical model to pass the point of temporary instability (negative stiffness) to reach a stable state. However, this approach is only applicable when there is a single point of loading.

Where there arc different loadings, the Riks method can be used as alternative. However, the Riks method can only be applied in the load–deflection domain. Therefore, for structural analysis in fire, it can only be used in steady-state analysis where the structural temperatures remain unchanged. An example of such implication can be found in the work by Jafarian and Wang ( 2015a,b).

In real scenarios, transient modelling is more often employed when simulating structural behaviour in fire, in which the external mechanical load on the structure remains constant while the structure experiences a change in temperature. To deal with temporary loss of stability, an effective method is to introduce artificial damping to enable the structure to jump from one state of stability before losing temporary stability to the other state of stability after recovering from temporary loss of stability. The damping factor should be carefully chosen (as small as possible) so that the viscous damping energy is a very small proportion of the potential energy of the system. Additionally, the reaction forces at any point should be checked to ensure that the damping force is negligible. An example of applying artificial damping to modelling connection behaviour in fire is Elsawaf et al. (2011), which suggests a value of 0.00001 as an appropriate damping factor for the simulation of a beam-to-column connection under fire conditions.

Sometimes it becomes necessary to simulate structural behaviour in fire using explicit dynamic analysis. How to scale the long fire exposure duration down to a short time without inducing an artificial dynamic effect is the key consideration. Examples of explicit dynamic simulation of structures in fire include Yu et al. (2008) and Albrifkani and Wang (2016).

### 2.4.3 Excessive computational time

Modelling the detailed behaviour of connections, which have many small components, within a large structure, can require a huge amount of computation time. It is worthwhile employing some techniques that can be used to shorten the computation time. An obvious technique is to model part of the structure, by taking advantage of symmetry (e.g. Beshir, 2016) and feasibility of separating the part of the structure of concern from the rest of the structure. A key consideration is to ensure that the boundary conditions to the part of the structure of concern are compatible with the expected force and deformation modes between the part being modelled and the rest of the structure.

A more effective approach, which can be used in conjunction with the above, or when modelling the whole structure, is to use a small number of elements to model the majority of the structure which is cold and whose behaviour can be approximated by simple frame analysis as undertaken in ambient temperature design.

For example, in Dai et al. (2010) and Elsawaf et al. (2011), as shown in Figure 2.9, adopted this approach. The connection and its immediate surrounding structure (connected beam and part of connected columns) are modelled in detail but the rest of the structure is represented by line elements.

Even the connection of concern can be simplified by using a component-based spring model to represent the connection of concern. This also helps minimise the problem of non-convergence caused by modelling interface.

Clearly, the accuracy of component-based modelling method hinges on correct identification of connection components and accurate quantification of their load–deflection curves at elevated temperatures. The component-based method for connection modelling is well established at

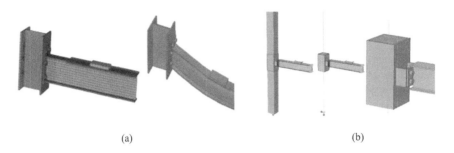

(a)                                                             (b)

Figure 2.9 Combination of solid elements with line elements, (a) Dai et al. (2010) and (b) Elsawaf et al. (2011).

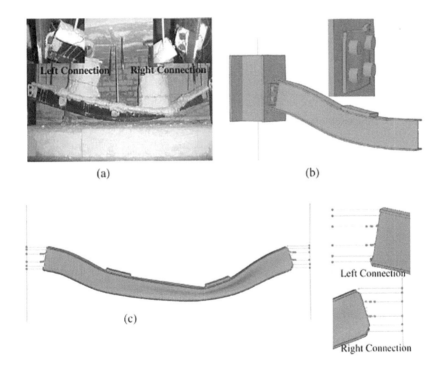

Figure 2.10 Example of incorporation of component-based model in frame analysis Chen and Wang (2012). (a) Test. (b) Detailed Model (LHS). (c) Hybride Model.

ambient temperature (EN1993-1-8 (CEN, 2005)) and is now extensively adapted to elevated temperature applications. Sources of information for quantifying connection component load–displacement curves and developing component-based models for connections at elevated temperatures include Quan et al. (2017), Liu et al. (2020) and Wald et al. (2020). Further details are presented in Chapter 4.

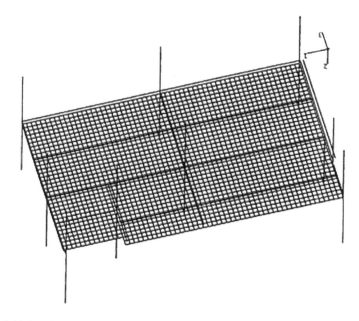

*Figure 2.11* Simulation of a Cardington fire test by Gillie et al. (2002).

An example of incorporating component-based modelling of connections in frame analysis is shown in Figure 2.10, based on Chen and Wang (2012).

Similarly, the floor slab can be modelled by using shell elements as shown in Figure 2.11.

## 2.5 SUMMARY

This chapter has outlined the complex behaviour, in terms of variations of internal forces and deflections, of steel connections as part of a whole structure where there are strong interactions between the connection and the rest of the structure. These complexities cannot be fully dealt with in any simple analytical method that may be employed in everyday design practice. Therefore, the main part of this chapter is devoted to identifying the connection forces at different key stages that should be quantified so that connection adequacy can be checked. This is then followed by presenting simple calculation methods with examples to demonstrate their use and their validity. The key connection quantities to be calculated are as follows: maximum compression force during heating (stage 1), critical temperature at bending resistance (stage 2), tensile and bending moment at catenary stage (stage 3) and residual tensile force at ambient temperature cooling from the critical temperature.

While the proposed simple methods should be able to cover the most common types of connections and structural layouts, there are some cases where a more detailed simulation of connection behaviour in fire may be required. The final section of this chapter outlines how a few of the most challenging problems of modelling connection and structural behaviour in fire may be tackled, including non-convergence problems caused by material failure and loss of temporary stability of the structure and extensive computation time.

## REFERENCES

Al-Jabri, K., Burgess, I., Lennon, T. & Plank, R. 2005. Moment–rotation–temperature curves for semi-rigid joints. *Journal of Constructional Steel Research*, 61, 281–303.

Al-Jabri, K. S. 1999. *The Behaviour of Steel and Composite Beam-to-Column Connections in Fire*. University of Sheffield, Sheffield, UK.

Albrifkani, S. & Wang, Y. C. 2016. Explicit modelling of large deflection behaviour of restrained reinforced concrete beams in fire. *Engineering Structures*, 121, 97–119.

Beshir, M. A. E. Z. 2016. *Robustness of Composite Framed Structures in Fire*, The University of Manchester (United Kingdom).

Brown, D., Iles, D., Brettle, M. & Malik, A. 2013. Joints in Steel Construction: Moment-Resisting Joints to Eurocode 3. *BCSA/SCI Connections Group*

CEN 2005. EN 1993-1-8 Eurocode 3: Design of steel structures, Part 1–8: Design of joints. *British Standard*.

Chen, L. & Wang, Y. C. 2012. Efficient modelling of large deflection behaviour of restrained steel structures with realistic endplate beam/column connections in fire. *Engineering Structures*, 43, 194–209.

Dai, X., Wang, Y. & Bailey, C. 2010. Numerical modelling of structural fire behaviour of restrained steel beam–column assemblies using typical joint types. *Engineering Structures*, 32, 2337–2351.

Ding, J. & Wang, Y. C. 2007. Experimental study of structural fire behaviour of steel beam to concrete filled tubular column assemblies with different types of joints. *Engineering Structures*, 29, 3485–3502.

Elsawaf, S., Wang, Y. & Mandal, P. 2011. Numerical modelling of restrained structural subassemblies of steel beam and CFT columns connected using reverse channels in fire. *Engineering Structures*, 33, 1217–1231.

Gillie, M., Usmani, A. & Rotter, J. M. 2002. A structural analysis of the Cardington British steel corner test. *Journal of Constructional Steel Research*, 58, 427–442.

Jafarian, M. & Wang, Y. 2015a. Tying resistance of reverse channel connection to concrete filled square and rectangular tubular sections. *Engineering Structures*, 100, 17–30.

Jafarian, M. & Wang, Y. C. 2015b. Force–deflection relationship of reverse channel connection web component subjected to transverse load. *Journal of Constructional Steel Research*, 104, 206–226.

Lawson, R. M. 1990. Behaviour of steel beam to column connections in fire. *Structural Engineering*, 68, 8.

Leston Jones, L., Jones, L., Burgess, I., Lennon, T., Plank, R. & BRE 1997. Elevated-temperature moment-rotation tests on steelwork connections. *Proceedings of the Institution of Civil Engineers-Structures and Buildings*, 122, 410–419.

Liu, T. C. H., Fahad, M. K. & Davies, J. M. 2002. Experimental investigation of behaviour of axially restrained steel beams in fire. *Journal of Constructional Steel Research*, 58, 1211–1230.

Liu, Y., Huang, S.-S. & Burgess, I. 2020. Performance of a novel ductile connection in steel-framed structures under fire conditions. *Journal of Constructional Steel Research*, 169, 106034.

Lopes, F., Santiago, A., Simões da Silva, L., Heistermann, T., Iqbal, N., Veljkovic, M., Dong, G., Huang, S., Davison, B., Burgess, I. W., Jafarian, M. & Wang, Y. C. 2011. COMPFIRE Design of composite joints for improved fire robustness.

Nethercot, D. A. 2006. Connection research and its impact on practice during the Dowling era. *Journal of Constructional Steel Research*, 62, 1165–1170.

Quan, G., Huang, S.-S. & Burgess, I. 2017. The behaviour and effects of beam-end buckling in fire using a component-based method. *Engineering Structures*, 139, 15–30.

Ramli-Sulong, N., Elghazouli, A. & Izzuddin, B. 2007. Behaviour and design of beam-to-column connections under fire conditions. *Fire Safety Journal*, 42, 437–451.

Usmani, A. S., Rotter, J. M., Lamont, S., Sanad, A. M. & Gillie, M. 2001. Fundamental principles of structural behaviour under thermal effects. *Fire Safety Journal*, 36, 721–744.

Wald, F., Šabatka, L., Bajer, M., Jehlička, P., Kabeláč, J., Kožich, M., Kuříková, M. & Vild, M. 2020. *Component-based Finite Element Design of Steel Connections*, Czech Technical University.

Wang, Y. C. 2002. *Steel and Composite Structures*. First ed. Taylor & Francis, London and New York.

Wang, Y. C., Dai, X. H. & Bailey, C. G. 2011. An experimental study of relative structural fire behaviour and robustness of different types of steel joint in restrained steel frames. *Journal of Constructional Steel Research*, 67, 1149–1163.

Yin, Y. 2004. Advanced behaviour of steel beams under fire conditions. University of Manchester.

Yin, Y. Z. & Wang, Y. C. 2004. A numerical study of large deflection behaviour of restrained steel beams at elevated temperatures. *Journal of Constructional Steel Research*, 60, 1029–1047.

Yu, H., Burgess, I., Davison, J. & Plank, R. 2008. Numerical simulation of bolted steel connections in fire using explicit dynamic analysis. *Journal of Constructional Steel Research*, 64, 515–525.

# Chapter 3

# Connection temperatures in fire

## 3.1 INTRODUCTION

After quantifying fire behaviour (which is not covered in this book), the next step of evaluating connection behaviour is to calculate connection temperatures. Because a connection consists of many components, the connection temperature distribution is complex. Therefore, for design purpose, simplified and safe methods of calculating connection temperatures are needed, which are covered in this chapter. As mentioned in Chapter 1, this book adopts the component-based method for quantifying connection behaviour. Therefore, the connection temperature calculation methods in this chapter will be for calculating temperatures of connection components. Furthermore, the temperature calculation equations will be the same as the equations in EN 1993-1-2 (CEN, 2005) for unprotected and protected steel members by using section factor. Calculating connection temperatures requires input of thermal properties of steel. For brevity, this information is not presented in this chapter. However, for completeness, the thermal properties of different types of steel, extracted from EN 1993-1-2 (CEN, 2005) are provided in Appendix A of this book. Therefore, the focus of this chapter is to present calculation methods for section factors of various connection components.

Based on the findings of a number of previous research investigations (Lawson, 1990, Steel, 1999, Al-Jabri, 1999, Wald et al., 2006, 2009, Zhao et al., 2007, Ding and Wang, 2009), the temperature distribution in unprotected connections may be assumed to be non-uniform with different connection components having different temperatures. However, the temperature distribution in protected connections may be assumed to be uniform in all connection components with the connection temperature equal to that of the connected beam. Therefore, this chapter will describe methods of calculating section factors for different connection components without fire protection and will also present supporting information to illustrate their accuracy and that they are safe to use.

There are situations where protected connections are connected to unprotected members. In these situations, the protected connection is not

DOI: 10.1201/9781003134466-3

only exposed to fire but also receives additional conductive heat from the unprotected steel member. To minimise the conductive heat, the length of the unprotected member has to be protected. This protection length is referred to as coatback. This chapter will present existing solutions to coatback length for a few applications.

In bolted connections, protecting the bolts presents some practical problems. One possible solution is to use a proprietary system called bolt cap.

As justification for the above two assumptions, Figure 3.1 presents measured connection temperature distributions and connected steel beam temperatures from a few representative fire tests. The results show that the temperature at the connection is considerably lower than the beam's lower flange measured at the mid-span of the arrangement. Also, at the connection position, the temperature would be reduced by moving away from the lower flange and getting closer to the composite slab. This confirms that assuming a uniform temperature distribution at the connection position would be a safe assumption.

## 3.2 LOCAL SECTION FACTORS FOR CONNECTION COMPONENTS

For unprotected steel, the general equation for calculating the steel temperature in Eurocode EN 1993-1-2 (CEN, 2005) is

$$\Delta\theta_{a,t} = k_{sh} \frac{A_m/V}{c_a\rho_a} \dot{h}_{net} \Delta t \tag{3.1}$$

where
$k_{sh}$ is the correction factor for shadow effect
$A_m/V$ is the section factor of the unprotected steel member $[m^{-1}]$
$A_m$ is the surface area of the section per unit length $[m^2/m]$
$V$ is the volume of the member per unit length $[m^3/m]$
$c_a$ is the specific heat of steel $[J/kg \cdot K]$
$\rho_a$ is the unit mass of steel $[kg/m^3]$

$\dot{h}_{net}$ is the design value of the net heat flux per unit area $[W/m^2]$
$\Delta\theta_{g,t}$ is the increase of the ambient gas temperature during the time interval $\Delta t$ $[K]$
$\Delta t$ is the time interval [Second]

$$\phi = \frac{c_p\rho_p}{c_a\rho_a} d_p A_p/V$$

For calculating temperatures of connection components in fire, local section factors can be used.

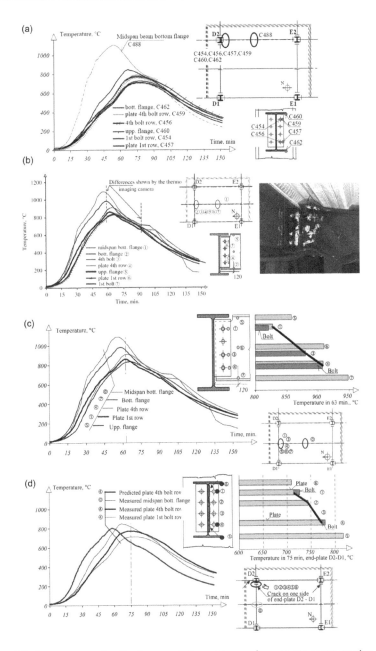

*Figure 3.1* Temperature distributions on different types of connections exposed to natural fire. (a) Beam and connection temperature distribution (Wald et al., 2006).
(b) Recorded temperature around fin plate connection and connection status
after the test (Wald et al., 2006). (c) Temperature variation on a fin plate connection (Wald et al., 2004). (d) Temperature distribution on a partial depth
endplate connection (Wald et al., 2006).

In the component-based method, bolts are separate components from the bolted plates. However, when calculating bolt temperatures, bolts are assumed to have the same temperature as the bolted plates. Bolts have lower section factors than the bolted plates, resulting in lower bolt temperatures. Therefore, assuming bolts have the same temperature as the bolted plate is safe.

Tables 3.1 and 3.2 list different common types of steel I-section beam to H-section column, steel I-section beam to concrete-filled tubular column connections, their components for calculating connection temperatures and equations for calculating their local section factors.

To demonstrate accuracy and safety of the methods in Tables 3.1 and 3.2, Figures 3.2 and 3.3 show comparison of connection component temperatures measured in fire tests with calculation results using the proposed connection component section factors in this chapter for the different types of connections in Tables 3.1 and 3.2.

## 3.3 NEW VERSION OF EN1993-1-2

As a further simplification, the beneficial effects of bolt in reducing the local section factor of the bolted steel plate may be safely ignored. Therefore, in the pre-standard version of new EN1993-1-2(CEN, 2021), the local section factor for bolted connection is that of a steel plate with a thickness of the total of the bolted steel plates. The simple equation to calculate the connection local section factor is

$$A_m/V = \frac{2}{t} \qquad (3.2)$$

where in the above $t$ is the total thickness of the connected steel plates (e.g. endplate/column flange, fin plate/beam web) in the thinnest part of the joint.

## 3.4 COATBACK

There are some situations where a protected connection is at the end of an unprotected member. Figure 3.4 shows the situation of unprotected secondary beams in tensile membrane action applications. To minimise conductive heat transfer to the protected perimeter beams, the connection and the length of the secondary beams should be protected.

Figure 3.5 shows a connection to an unprotected concrete-filled steel tubular column. The connection is protected. To ensure that any temperature rise in the connection component closest to the unprotected, concrete-filled, tubular column is minimised, the length of the concrete-filled tubular column is protected below the connection line.

*Table 3.1* Local connection component section factors for common types of I-section beam to H-section column connections (Dai et al., 2010)

| Connection | Component | Local section factor |
|---|---|---|
| Fin plate | | $$\frac{A_m}{V} = \frac{2\left(L_f + t_f + t_w\right)}{L_f\left(t_f + t_w\right)}$$ |
| Web cleat | | $$\frac{A_m}{V} = \frac{2\left(L_{cl} + t_{cf} + W_{cf}\right)}{\left(L_{cl} - t_c\right)\left(2\,t_c + t_w\right) + 2L_{c2}t_c + W_{cf}t_{cf}}$$ |
| Endplate | | $$\frac{A_m}{V} = \frac{2\left(W_{cf} + t_{cf} + t_e\right)}{W_{cf} \times t_{cf} + t_e \times W_e}$$ |

$L_f$ is the fin plate length (m),
$t_f$ is the fin plate thickness (m),
$t_w$ is the beam web thickness (m),
$L_{cl}$ is the web cleat length in contact with beam (m),
$L_{c2}$ is the web cleat length in contact with column (m).

$t_{cf}$ is the column flange thickness (m),
$W_{cf}$ is the column flange width (m),
$t_c$ is web cleat thickness (m),
$W_e$ is the endplate width (m),
$t_e$ is the endplate thickness (m).

*Table 3.2* Local section factors for I-section to concrete-filled hollow section connections (Ding and Wang, 2009)

| | | |
|---|---|---|
| **Fin plate** | 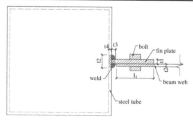<br>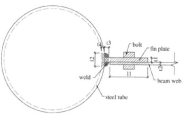 | Fin plate in contact with the beame<br><br>$$\frac{A_m}{V} = \frac{2}{(t_1 + t_2)}$$<br><br>Fin plate in contact with the column<br><br>$$\frac{A_m}{V} = \frac{\pi t_3 + t_1}{\dfrac{\pi t_3{}^2}{2} + t_1 t_3 + t_4 l_2}$$ |
| **Endplate** | | Endplate in contact with the beam<br><br>$$\frac{A_m}{V} = \frac{1}{(t_1 l_1 + t_2 l_2)}$$<br><br>Endplate in contact with the column<br><br>$$\frac{A_m}{V} = \frac{\pi t_3 + t_4}{\dfrac{\pi t_3{}^2}{2} + t_3 t_4 + (t_1 + t_2) l_3}$$ |
| **T-Stub** | | T-Stub in contact with the beam<br><br>$$\frac{A_m}{V} = \frac{2}{(t_3 + t_4)}$$<br><br>T-Stub in contact with the column<br><br>$$\frac{A_m}{V} = \frac{1}{(t_1 l_1 + t_2 l_2)}$$ |
| **Reverse channel** | 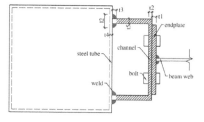 | Reverse channel in contact with beam<br><br>$$\frac{A_m}{V} = \frac{2}{(t_1 + t_2)}$$<br><br>Reverse channel in contact with the column<br><br>$$\frac{A_m}{V} = \frac{\pi t_3 + t_4}{\dfrac{\pi t_3{}^2}{2} + t_3 t_5 + t_4 l_3}$$ |

For the application shown in Figure 3.6, two methods exist to specify the coatback length. In the United Kingdom, the ASFP technical guidance document 8 (ASFP, 2010) specifies a value of 500 mm as the minimum. However, this value may be reduced if supported by fire test evidence from the fire protection manufacturer.

Additionally, for applications under hydrocarbon fire, FABIG (Fire And Blast Information Group, administered by the UK's Steel Construction Institute) note 13 (SCI, 2014), specifies a coatback length of 450 mm. This value is based on numerical simulation results but was confirmed by the independent study of Friebe et al. (2014).

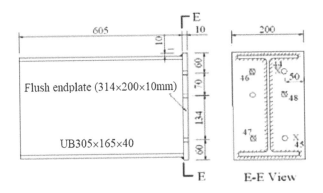

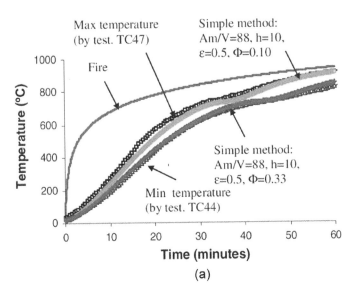

(a)

*Figure 3.2* Comparison between temperatures calculated using suggested section factors and test data (Dai et al., 2010).

*(Continued)*

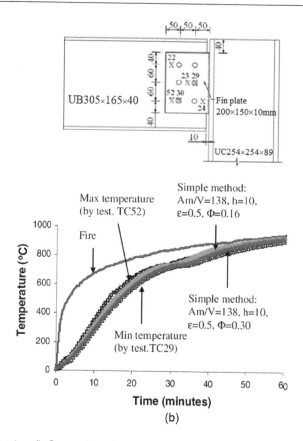

Figure 3.2 *(Continued)* Comparison between temperatures calculated using suggested section factors and test data (Dai et al., 2010).

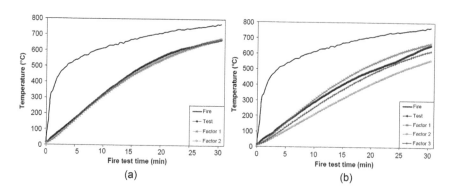

Figure 3.3 Comparison between temperatures calculated using suggested section factors and test data (Ding and Wang, 2009). (Note: in the above graphs, Factor I in Figure (a) and Factor 3 in Figure (b) represent the formulations given in Table 3.2.)

Figure 3.4 Unprotected secondary beam with coatback at the connection (Li et al., 2017, Dai et al., 2009).

Figure 3.5 Example of coatback for a connection to concrete-filled hollow section (Wang, 2014).

In Germany, DIN 4102-4:2016 (DIN, 2016) suggests a value of 300 mm for standard fire resistance periods up to 90 minutes and 600 mm for standard fire resistance periods of 120 minutes and above.

Specifying the coatback length as a function of fire resistance period is more logical. However, Both the UK and German specifications are approximate, with the UK value tending to be at the upper bound of the German value. Furthermore, any change for different fire resistance periods is relatively small. Therefore, either recommendation can be used in the absence of further refinement.

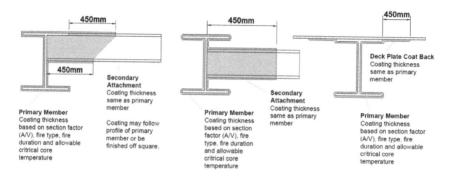

*Figure 3.6* Application of coatback when exposed to hydrocarbon fire (SCI, 2014).

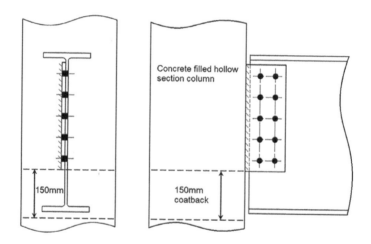

*Figure 3.7* Coatback on concrete-filled hollow sections.

For connections to unprotected, concrete-filled, tubular columns, Wang (2014), based on numerical simulation results, suggests a coatback of 150 mm, as shown in Figure 3.7.

## 3.5 WORKED EXAMPLES

1. Extended endplate connection as detailed in Figure 3.8.
    I.  Local section factor
        Using the method in Table 3.1, the local section factor for an end-plate connection can be calculated as follows

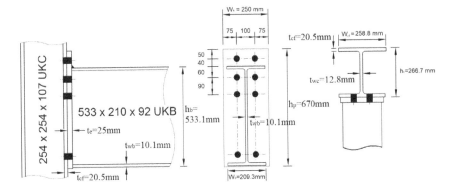

*Figure 3.8* Details of an endplate connection.

$$\frac{A_m}{V} = \frac{2\left(W_{cf} + t_{cf} + t_e\right)}{W_{cf} \times t_{cf} + t_e \times W_e}$$

$$= \frac{2 \times (258.8 + 20.5 + 25) \times 10^{-3}}{(258.8 \times 20.5 + 25 \times 250) \times 10^{-6}} = 52.67 \text{ m}^{-1}$$

II. New simplified method to EN1993-1-2

$A_m/V = 2/t$

$t = t_{cf} + t_e = 20.5 + 25 = 45.5 \text{ mm} = 0.0455 \text{ m}$

$A_m/V = 2/t = 2/0.0455 = 43.96 \text{ m}^{-1}$

2. Fin plate connection as detailed in Figure 3.9
   I. Local section factor
      Using the method in Table 3.1, the local section factor for a fin
      plate connection can be calculated as follows:

$$\frac{A_m}{V} = \frac{2\left(L_f + t_f + t_w\right)}{L_f\left(t_f + t_w\right)} = \frac{2 \times (150 + 10 + 10.1) \times 10^{-3}}{150 \times (10 + 10.1) \times 10^{-6}} = 112.84 \text{ m}^{-1}$$

   II. New simplified method to EN1993-1-2

$A_m/V = 2/t$

$t = t_f + t_w = 10.1 + 10 = 20.1 \text{ mm} = 0.0201 \text{ m}$

$A_m/V = 2/t = 2/0.0201 = 99.50 \text{ m}^{-1}$

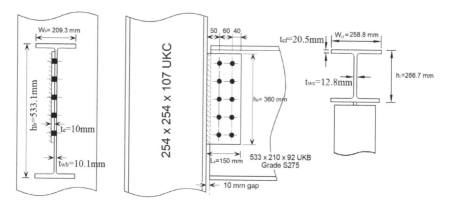

*Figure 3.9* Details of a fin plate connection.

## REFERENCES

Al-Jabri, K. S. 1999. *The behaviour of steel and composite beam-to-column connections in fire.* University of Sheffield.

ASFP 2010. Technical Guidance Document 8: Code of practice for junctions between different fire protection systems when applied to load bearing structural steel elements. ASFP.

CEN 2005. EN 1993-1-2 Eurocode 3: Design of Steel Structures–Part 1–2: General Rules–Structural Fire Design. *British Standards Institution.*

CEN 2021. prEN 1993-1-2 Eurocode 3: Design of Steel Structures–Part 1–2: General Rules–Structural Fire Design. *British Standards Institution.*

Dai, X., Wang, Y.-C. & Bailey, C. 2009. Effects of partial fire protection on temperature developments in steel joints protected by intumescent coating. *Fire Safety Journal*, 44, 376–386.

Dai, X., Wang, Y. & Bailey, C. 2010. A simple method to predict temperatures in steel joints with partial intumescent coating fire protection. *Fire Technology*, 46, 19.

DIN 2016. *DIN 4102-4. Fire behaviour of building materials and building components - Part 4: Synopsis and application of classified building materials, components and special components.* Deutsches Institut fur Normung E.V. (DIN), Berlin.

Ding, J. & Wang, Y. C. 2009. Temperatures in unprotected joints between steel beams and concrete-filled tubular columns in fire. *Fire Safety Journal* 44, 16–32.

Friebe, M., Jang, B.-S. & Jim, Y. 2014. A parametric study on the use of passive fire protection in FPSO topside module. *International Journal of Naval Architecture and Ocean Engineering*, 6, 826–839.

Lawson, R. 1990. *Enhancement of fire resistance of beams by beam to column connections*, Steel Construction Institute UK.

Li, G.-Q., Zhang, N. & Jiang, J. 2017. Experimental investigation on thermal and mechanical behaviour of composite floors exposed to standard fire. *Fire Safety Journal*, 89, 63–76.

SCI 2014. Design guidance for hydrocarbon fires/Steel Construction Institute. *FABIG Technical Note; 13*. Steel Construction Institute, London.

Steel, B. 1999. The behaviour of multi-storey steel framed buildings in fire. *British Steel, Rotherham, UK*, 82.

Wald, F., Da Silva, L. S., Moore, D., Lennon, T., Chladna, M., Santiago, A., Beneš, M. & Borges, L. 2006. Experimental behaviour of a steel structure under natural fire. *Fire Safety Journal*, 41, 509–522.

Wald, F., Silva, S., Moore, D. & Lennon, T. 2004. Structural integrity fire test. *Proceedings Nordic Steel Conference*.

Wald, F., Sokol, Z. & Moore, D. 2009. Horizontal forces in steel structures tested in fire. *Journal of Constructional Steel Research*, 65, 1896–1903.

Wang, Y. 2014. *Design guide for concrete filled hot finished structural hollow section (SHS) columns*, TATA Steel.

Zhao, B., Roosefid, M., Breunese, A., Koutlas, G., Zilli, G., Hanus, F. & Franssen, J. 2007. Connections of steel and composite structures under natural fire conditions. *Contract number RFSR-CT-2006-00028, RFCS Mid-term Report*.

# Chapter 4

# Connection components and their force–displacement relations

## 4.1 INTRODUCTION

Extensive research studies on connection behaviour at ambient temperature date back more than 50 years (Nethercot, 2006), the main focus being to exploit the benefits of semi-rigid and partial-strength capabilities of conventional connections under bending in structural engineering design in cold conditions. Many physical tests were carried out, and the test results were collated to establish a database of connection moment-rotation curves to enable designers and researchers to make use of the test results in structural engineering research and design. However, whilst being able to directly use moment-rotation curves of connections makes it easy for research and design work, no result of connection moment-rotation curve would be available if the user wants to make some changes to some aspects of the connections (e.g. dimensions and positions of bolts) that are not in the database. Recognising that all conventional connections are made of a limited number of types of components, albeit with different positions and dimensions, it is possible that by thoroughly understanding the behaviour of these components under different conditions, they can then be assembled to predict connection moment-rotation curves for any variations in component positions and dimensions. This is the concept of the component-based method that is now firmly established as the basis of quantifying connection behaviour, as enshrined in EN1993-1-8 (CEN, 2005).

In the component-based method, the connection is an assembly of a series of non-linear spring elements each representing an active component of the connection. The force–displacement relationship of each spring element is the force–displacement curve of the active component. Under a rotation and movement in the longitudinal direction of the connected beam, axial displacements of all the springs are calculated relative to the centre of rotation of the connection. From their displacements, forces in all the springs can be calculated. The bending moment of the connection can then be calculated as the sum of contributions from all active springs.

Although the component-based method was developed for structural engineering research and design in cold conditions, its flexibility is ideally

DOI: 10.1201/9781003134466-4

suited to applications in fire. As explained in detail in Chapter 2, under fire condition, the forces in connections are not only bending moments but also axial forces. Furthermore, these forces vary with time and temperature. By quantifying the effects of temperature on component force–displacement relationships, the component-based method can be readily adapted for applications in fire, as demonstrated by Spyrou et al. (2002), Hu et al. (2009), Yu et al. (2009b), Chen and Wang (2012) and Taib and Burgess (2013). This book will use detailed examples to demonstrate how to use the component-based method to check connection resistance in fire.

Implementation of the component-based method in fire involves the following steps:

- Representing the connection by a series of springs, either in parallel or in series. This is part of the presentation of this chapter.
- Quantifying temperatures of all the springs at any particular time in fire. This has been dealt with in Chapter 3.
- Determining force–displacement relations of all the springs at elevated temperatures.
- Determining the overall connection axial deformation and rotation at the fire exposure time, as well as total resultant axial force and bending moment in the connection, based on the behaviour of the connected beam. Chapter 2 has presented details on these.
- Calculating connection resistances to overall axial force and bending moment that are compatible with the overall deformation, by assembling contributions of all springs (connection components), and comparing them with the above.

This chapter will identify active and independent connection components that are present in all common types of steel and composite connections and present their force–displacement relations at elevated temperatures. Chapter 5 will use examples to show detailed implementation of the component-based method.

## 4.2 IDENTIFICATION OF COMPONENTS FOR COMMON TYPES OF CONNECTIONS

This chapter deals with the following common types of steel and composite connection and their corresponding tables list their active connection components:

- Fin plate connection (Table 4.1a)
- Endplate connection (Table 4.1b)
- Web cleat connection (Table 4.1c)

It should be mentioned that the shear force in a connected beam in fire usually does not change in fire and acts independently of the bending moment and axial force in the beam which vary in fire and interact with each other. Checking for shear resistance of the connection is relatively straightforward. Therefore, although checking for shear resistance of connections is needed in fire resistance design of connections, application of the component-based method in this book will only focus on combined bending and axial force in the connected beam.

As a summary of Table 4.1, the following is a complete list of active and independent connection components from which the above-mentioned three common types of connection can be constructed:

    i. T-Stub model – column flange in bending
   ii. T-Stub model – endplate in bending
  iii. T-Stub model – bolt in tension
   iv. Column web in tension
    v. Plate in tension or compression
   vi. Plate in bearing
  vii. Bolt in shear
 viii. Weld in tension
   ix. Column web in compression
    x. Beam flange in compression

These components are the same for both steel and composite connections. For composite connections, the contribution from reinforcement should be taken into account according to EN 1994-1-1 (CEN, 2004).

## 4.3 FORCE–DISPLACEMENT RELATIONS OF ACTIVE CONNECTION COMPONENTS

Typically, the force–displacement relation of any active connection component is nonlinear, as sketched in Figure 4.1. For simplification, this curve is represented by a bilinear elastic/elastic–plastic relation as shown in Figure 4.1, which in general would be suitable for most of the components. However, for some components such as T-stub and plate in bearing, a different force–displacement curve representation is needed, and this will be described in relevant sub-sections.

Table 4.2 summarises the sources of information for these quantities for all the connection components listed at the end of the previous section. The main source of information is EN 1993-1-8 for ambient temperature design unless further research studies at elevated temperatures have suggested better alternatives which will be identified. When adapting ambient temperature solutions for elevated temperature applications, for the initial stiffness

*Table 4.1* Construction of common types of connection using connection components

a. Connection type: Fin plate

- Fin plate in bearing (vi),
- Beam web in bearing (vi)
- Bolt in shear (vii)
- Beam flange under compression [(1)](x)
- Column flange under compression [(1)](ix)

[(1)] active only when the beam and column flanges contact at large rotations, as shown in Figure 4.6.

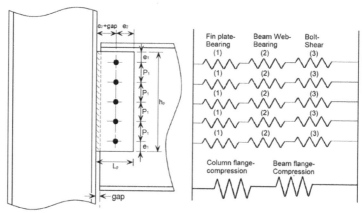

b. Connection type: Flexible/flush/extended endplate connection

- Column flange in bending (i)
- Column web in tension (ii)
- Endplate in bending (v)
- Endplate in bearing (iv)
- Bolt in tension (vii)
- Bolt in shear (viii)
- Weld in tension (ix)
- Column web in compression (x)
- Beam flange in compression (xi)

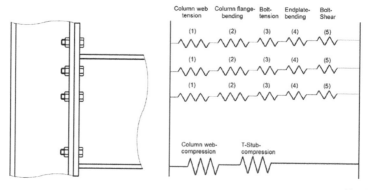

*(Continued)*

*Table 4.1 (Continued)*   Construction of common types of connection using connection components

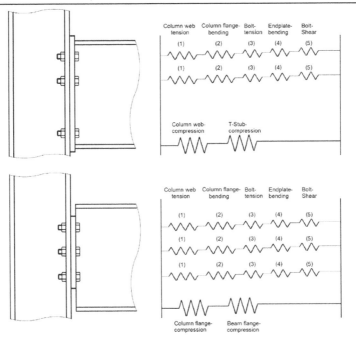

c. Connection type: Web cleat

- Column flange in bending (i)
- Column web in tension (ii)
- Angle leg in bending (v)
- Angle leg in bearing (iv)
- Beam web in tension (iii)
- Bolt in tension (vii)
- Bolt in shear (viii)
- Weld in tension (ix)
- Angel flange (plate) in bearing (iv)
- Beam web in bearing (iv)
- The bolt in shear (viii)
- Column web in compression [4] (x)
- Beam flange in compression [4] (xi)

[4] active only when beam flange and column flange contact at large rotations.

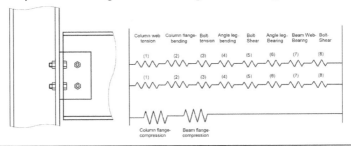

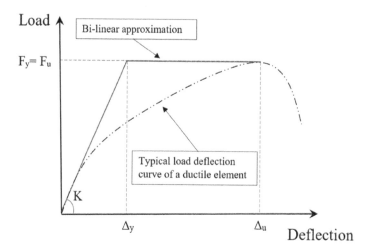

*Figure 4.1* Typical and simplified force–deflection curve of a connection component.

and resistance, these quantities will be modified by multiplying the ambient temperature values by the reduction factors for Young's modulus and yield stress of steel and bolts at elevated temperatures respectively. For deflection capacity, Eurocode EN 1993-1-8 (CEN, 2005) does not offer any solution. Therefore, the deflection capacities of the identified connection components are based on the outcome of research studies in literature. These sources of information are what the authors believe to be the most reliable information at the time of writing. However, future research studies may suggest improved formulations. Nevertheless, the methodology of quantifying and checking the connection behaviour of this book should still be applicable.

The remaining part of this chapter presents detailed calculation equations for the quantities listed in Table 4.2. To avoid repeating the information given in EN1993-1-8 (CEN, 2005), this section focuses on the quantities that are not using the standard, which are indicated by symbol[2] in Table 4.2.

In all the following equations, the mechanical properties (Young's modulus, yield stress, ultimate tensile stress) are those at elevated temperatures.

## 4.3.1 T-Stub model Figure

In research studies on T-stub (e.g. column flange or endplate), the assembly of a T-stub and its connected bolts is used. Therefore, the totality of T-stub and bolt displacements is obtained. However, since the bolt connects two T-stubs, it cannot be subsumed into any of the connected T-stub components, and the bolt must be treated as an independent component. Therefore, the displacements of the T-stub and the bolt must be separated from the totality.

Table 4.2 Connection components and sources of information for different quantities of their force–displacement relations

| Connection component | Sources of equations to calculate component force–displacement relations | | |
| --- | --- | --- | --- |
| | Resistance | Stiffness coefficient | Deformation capacity |
| i   T-Stub model –column flange in bending | T-Stub model section – EN 1993-1-8 (CEN, 2005)[a] | T-Stub model section – EN 1993-1-8 (CEN, 2005) | Spyrou et al. (2004)[b] |
| ii   T-Stub model –endplate in bending | T-Stub model section – EN 1993-1-8 (CEN, 2005)[a] | T-Stub model section – EN 1993-1-8 (CEN, 2005) | Spyrou et al. (2004)[b] |
| iii   T-Stub model –bolts in tension | EN 1993-1-8 (CEN, 2005)[a] | Chen and Wang (2012)[b] | Hu et al. (2009)[b] |
| iv   Column web in transverse tension | EN 1993-1-8 (CEN, 2005) | EN 1993-1-8 (CEN, 2005) | Beg et al. (2004)[b] |

(Continued)

Table 4.2 (Continued) Connection components and sources of information for different quantities of their force–displacement relations

| | Connection component | Sources of equations to calculate component force–displacement relations | | |
| --- | --- | --- | --- | --- |
| | | Resistance | Stiffness coefficient | Deformation capacity |
| v | Plate in tension or compression | EN 1993-1-8 (CEN, 2005) | EN 1993-1-8 (CEN, 2005) | EN 1993-1-8 (CEN, 2005) |
| vi | Plate in bearing | EN 1993-1-8 (CEN, 2005) | Sarraj (2007)[b] | Sarraj (2007)[b] |
| vii | Bolts in shear | EN 1993-1-8 (CEN, 2005) | Sarraj (2007)[b] | Sarraj (2007)[b] |
| viii | Welds | EN 1993-1-8 (CEN, 2005) | Hu et al. (2009)[b] | Hu et al. (2009)[b] |

(Continued)

Table 4.2 (Continued) Connection components and sources of information for different quantities of their force–displacement relations

| Connection component | Sources of equations to calculate component force–displacement relations | | |
|---|---|---|---|
| | Resistance | Stiffness coefficient | Deformation capacity |
| ix | Column web in transverse compression | EN 1993-1-8 (CEN, 2005) | EN 1993-1-8 (CEN, 2005) and Block (2006)[b] | Block (2006)[b] |
| x | Beam or column flange and web in compression | EN 1993-1-8 (CEN, 2005) | Infinity (or very high value) | – |

[a] Only for the T-stub model where the yield ($F_y$) and ultimate resistance of the component ($F_u$) are different. In other cases, it is assumed that the ultimate resistance of the component is the same as the yield resistance of the component (i.e. $F_u = F_y$).

[b] Denoting equations to be presented in Section 4.3 of this chapter.

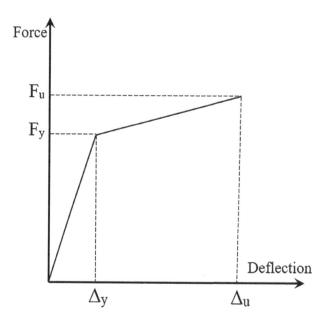

*Figure 4.2* Force–displacement behaviour of a T-stub in Mode I and Mode 2 failure.

In an equivalent T-stub representing the assembly of column flange/or endplate in bending, and bolt in tension, depending on the bending resistance of the column flange or endplate relative to the tensile resistance of the bolt, there are three failure modes as shown in 4.5: Mode 1 – formation of four plastic hinges in the plate (column flange or endplate); Mode 2 – formation of two plastic hinges in the plate and bolt failure; Mode 3 – bolt fracture. However, since the bolt is an independent component, the failure model of the T-stub (either flange or endplate without a bolt) is either Mode 1 or Mode 2.

For Mode 1 and Mode 2 failure of a T-sub, the force–displacement curve of the T-stub cannot be simplified by a linear elastic–plastic representation as shown in Figure 4.1. Instead, the yield resistance $(F_y)$ and the ultimate resistance $(F_u)$ have to be differentiated, as shown in Figure 4.2.

*Initial stiffness and yield resistance*: as given in EN1993-1-8 (CEN, 2005) with modifications for material properties of steel at elevated temperatures, for all failure modes.

*Ultimate resistance and deformation capacity*
*Failure Mode 1 (formation of four plastic hinges)*
    Adapted from Spyrou et al. (2004):

$$\Delta_u = \Delta_y + \Delta\delta_{cl} \tag{4.1a}$$

$$F_u = F_y + \Delta F = \frac{2M_p\left[2n + \dfrac{7k}{8}\right]}{mn + \dfrac{3km}{8} + \dfrac{3kn}{8} + \dfrac{k^2}{8}}$$  (4.1b)

where $\Delta_y = \dfrac{F_y}{K}$ and

$$\Delta\delta_{cl} = \frac{\Delta F}{E_t I}\left[\frac{\left(m + \dfrac{k}{4}\right)^2\left(m + \dfrac{k}{2}\right)}{8} - \frac{\left(m + \dfrac{k}{4}\right)^3}{24} + \frac{k^3}{1536}\right]$$  (4.1c)

*Failure Mode 2 (formation of two plastic hinges and bolt fracture)*

$$\Delta_u = \Delta_y + \Delta\delta_{cl}$$  (4.2a)

$$F_u = F_y + \Delta F = \frac{2M_p + 4A_s f_{by}\left(n + \dfrac{k}{2}\right)}{n + k + m}$$  (4.2b)

where $\Delta_y = \dfrac{F_y}{K}$ and

$$\Delta\delta_{cl} = \frac{\Delta F}{EI}\left[\frac{\left(m + \dfrac{k}{4}\right)^2\left(m + \dfrac{k}{2}\right)}{8} - \frac{\left(m + \dfrac{k}{4}\right)^3}{24} + \frac{k^3}{1536}\right]$$  (4.2c)

It should be pointed out that in Spyrou et al. (2004), the expressions for the deformation capacity of T-stubs contain displacement of the bolt because they are for the equivalent T-stub consisting of both the T-stub and bolts. However, as explained at the start of this section, because the bolt is treated as an independent component, its displacement is removed from that of the equivalent T-stub.

*Mode 3 (bolt fracture)*
*Stiffness of bolts under tension:*
as given by Sarraj (2007) with the modification of material properties of steel for elevated temperatures:

$$K = \frac{E_b}{\dfrac{L_s}{A_b} + \dfrac{L_b}{A_s}} \qquad (4.3)$$

$$K = \frac{A_s E_b}{L_b} \left( \text{for fully threaded bolts} \right) \qquad (4.4)$$

*Deformation capacity:*
According to (Yu et al., 2009a):

$$\Delta_y = \frac{F_y}{K} \qquad (4.5)$$

$$\Delta_u = \varepsilon_{u,\,b} \left( L_b + \frac{2}{n_{th}} \right) \qquad (4.6)$$

For definition of symbols used for all three failure modes:
$E$ is the Young's modulus of the plate
$E_b$ is the Young's modulus of the bolt
$E_{bt}$ is the 1% of the Young's modulus of the bolt
$E_t$ is the 1.5% of Young's modulus of the plate
$A_s$ is the shank area of the bolt
$A_b$ is the nominal area of the bolt shank
$I = 2 L_{\text{eff}}\, t_f^3 / 12$
$L_b$ is the effective length of the bolt
$n_{th}$ is the number of threads per unit length of the bolt
$\varepsilon_{u,\,b}$ is the ultimate strain of the bolt
The dimensional parameters $(m, n, k, L_e, L_{\text{eff}}, t_f)$ are shown in Figure 4.3.

## 4.3.2 Column web in tension (iv)

*Stiffness and resistance:* as given in EN 1993-1-8 (CEN, 2005) with the modification of material properties of steel for elevated temperatures.
   *Deformation capacity:* according to Beg et al. (2004) as follows (for dimensions see Figure 4.4):

$$\Delta_u = \varepsilon_u d_{wc} \qquad (4.7)$$

$\varepsilon_u$ is the ultimate strain of the column web.

## 4.3.3 Plate in tension or compression (v)

*Resistance, stiffness, deformation capacity:* according to EN 1993-1-8 (CEN, 2005).

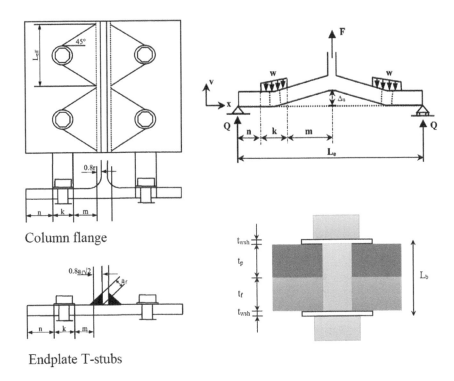

Column flange

Endplate T-stubs

*Figure 4.3* Dimensions of T-stub component (Spyrou et al., 2004).

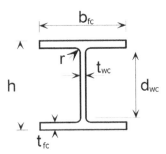

*Figure 4.4* Column dimensions.

## 4.3.4 Plate in bearing (vi)

For this component, the simplified representation in Figure 4.1 is not suitable. Instead, it is as shown in Figure 4.5. The following values are needed to establish the representative curve in Figure 4.5: ultimate bearing resistance of the plate $(F_u)$, initial stiffness of the plate (K), and ultimate deflection $(\delta_u)$.

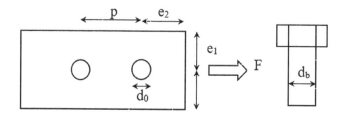

*Figure 4.5* Details of a plate in bearing.

*Initial stiffness*
In accordance with Sarraj (2007):

$$K = \cfrac{1}{\cfrac{1}{K_{br}} + \cfrac{1}{K_b} + \cfrac{1}{K_v}}$$

$$K_{br} = \Omega f_y t \left(\frac{d_b}{25.4}\right)^{0.8} \tag{4.8}$$

$$K_b = 32Et\left(\frac{e_2}{d_b} - 0.5\right)^3 \tag{4.9}$$

$$K_v = 6.67Gt\left(\frac{e_2}{d_b} - 0.5\right) \tag{4.10}$$

where
  $K_{br}$ is the bearing stiffness
  $K_b$ is the bending stiffness
  $K_v$ is the shearing stiffness
  $\Omega$, $\psi$ and $\phi$ are curve-fitting parameters and are given in Table 4.3 with reference to dimensions given in Figure 4.5.
  *Resistance*
  According to EN 1993-1-8 (CEN, 2005) with modifications for material properties.
  *Deformation capacity*
  Under compression, the component is assumed to have infinite deformation capacity (Figure 4.7).
  Under tension, the deflection at zero resistance ($\delta_u$) is the distance between the centre of the bolt hole to the edge of the plate in the direction of loading (shown as $e_2$ in Figure 4.6, indicating complete shearing out of the material in front of the bolt).

*Table 4.3* Parameters for bearing of a fin plate (Sarraj, 2007)

| T(°C) | $e_2 \leq 2d_b$ | | | $e_2 \geq 3d_b, d_b \leq 20$ | | | $e_2 \geq 3d_b, d_b = 24$ | | |
|---|---|---|---|---|---|---|---|---|---|
| | $\Omega$ | $\psi$ | $\Phi$ | $\Omega$ | $\psi$ | $\Phi$ | $\Omega$ | $\psi$ | $\Phi$ |
| 20 | 145 | 2.1 | 0.012 | 250 | 1.7 | 0.008 | 250 | 1.7 | 0.011 |
| 100 | 180 | 2 | 0.008 | 220 | 1.7 | 0.008 | 250 | 1.7 | 0.011 |
| 200 | 180 | 2 | 0.008 | 220 | 1.7 | 0.008 | 250 | 1.7 | 0.011 |
| 300 | 180 | 2 | 0.008 | 220 | 1.7 | 0.008 | 250 | 1.7 | 0.011 |
| 400 | 170 | 2 | 0.008 | 200 | 1.7 | 0.008 | 200 | 1.7 | 0.009 |
| 500 | 130 | 2 | 0.008 | 170 | 1.7 | 0.008 | 170 | 1.7 | 0.007 |
| 600 | 80 | 2 | 0.008 | 110 | 1.7 | 0.008 | 110 | 1.7 | 0.0055 |
| 700 | 45 | 2 | 0.008 | 40 | 1.7 | 0.007 | 40 | 1.7 | 0.0055 |
| 800 | 20 | 1.8 | 0.008 | 20 | 1.7 | 0.007 | 20 | 1.7 | 0.001 |

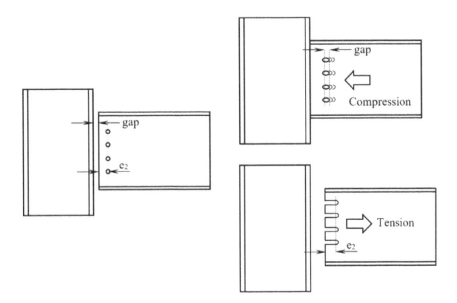

*Figure 4.6* Movement of the bolt in holes.

## 4.3.5 Bolt in shear (vii)

The idealised behaviour of a bolt in shear is shown in Figure 4.8, adapted from Sarraj (2007). The necessary information to construct such a curve is presented below.

*Initial stiffness:*

$$K = \frac{kA_sE_b}{2(1+v)d_b} = \frac{0.15GA_s}{d_b} \tag{4.11}$$

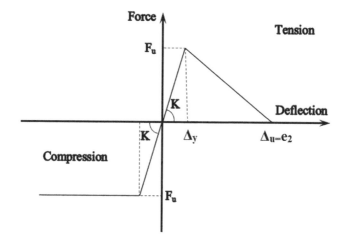

*Figure 4.7* Load–deflection curve for fin plate and beam web in bearing.

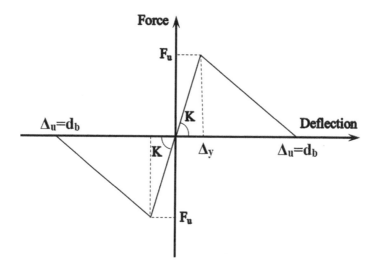

*Figure 4.8* Load–deflection curve for bolt under shear.

where
  $A_s$ is the cross-section area of the bolt
  $d_b$ is the diameter of the bolt
  $k$ is the shear correction factor recommended to be taken as 0.15
  $G$ is the temperature-dependent shear modulus of steel
  *Ultimate resistance:*

$$F_u = R_{f,v,b} f_{ub} A_s \tag{41.2}$$

where

$f_{ub}$ is the ultimate strength of the bolt

$R_{f,v,b} = k_{b,\theta}\,\alpha_v$ where $\alpha_v$ can be taken as 0.6 and $k_{b,\theta}$ is the strength reduction factor for bolt in shear as given in Eurocode 1993-1-8 (CEN, 2005).

*Deformation capacity:*

Based on the work by Yu et al. (2009c), $\delta_u$ (under shear) $= d_b$ (bolt diameter).

## 4.3.6 Welds (viii)

*Stiffness: infinite*

*Resistance:* according to EN 1993-1-8 with modification for elevated temperature.

*Deformation capacity:* 20% of the effective weld throat (Hu et al., 2009).

## 4.3.7 Column web in compression (ix)

*Stiffness:* As follows, according to Block (2006) with reference to Figure 4.9 for dimensions:

$$K = 0.95E\,\sqrt[4]{\frac{b_{fc}t_{wc}{}^2 t_{fc}{}^3}{b_{eff}d_{wc}}} \tag{4.13}$$

$$b_{eff} = t_{fb} + \sqrt{2}\,a_{ep} + t_{ep} + 5\left(t_{fc}+s\right) + \min\left(u; \sqrt{2}\,a_{ep}+t_{ep}\right)$$

where $b_{eff}$ is the effective length.

*Resistance:* with reference to the dimensions shown in Figure 4.9:

$$F_y = f_{y,c}\,t_{wc}\,\chi_F l_y \tag{4.14}$$

$$\chi_F = 0.06 + \frac{0.47}{\lambda_F} \le 1.0 \tag{4.15}$$

$$\lambda_F = \sqrt{\frac{l_y t_{wc} f_{y,wc}}{F_{cr}}}$$

$$F_{cr} = k_F\,\frac{\pi^2 E t_{wc}{}^3}{12\left(1-v^2\right)d_{wc}} \cong 0.9 k_F E\,\frac{t_{wc}{}^3}{d_{wc}}$$

$$l_y = c + 2t_{fc}\left(1+\sqrt{m_1+m_2}\right)$$

$$m_1 = \frac{f_{y,c}\,b_{fc}}{f_{y,wc}\,t_{wc}}$$

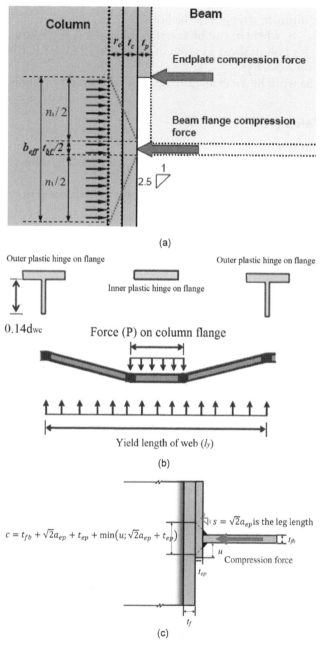

*Figure 4.9* Illustration of column web under compression. (a) Dimensions for column web in compression component (Hu et al., 2009). (b) Plastic hinge mechanism of column under concentrated loading (Block, 2006). (c) Load transfer between beam flange to column web.

$$m_2 = 0.02 \left( \frac{d_{wc}}{t_{fc}} \right)^2 \text{ if } \lambda_F < 0.05 \text{ then } m_2 = 0.0$$

$$c = t_{fb} + \sqrt{2} \; a_{ep} + t_{ep} + \min\left( u; \sqrt{2} \; a_{ep} + t_{ep} \right)$$

For the remaining parameters, refer to
   Figure 4.9 for dimensions.
   $\chi_F$ is the strength reduction factor for the web in compression
   $\upsilon = 0.3$
   $k_F = 3.5 + 2 \left( \dfrac{d_{wc}}{a_{wc}} \right)^2$ is the buckling coefficient for internal connections

   $k_F = 6 + 2 \left( \dfrac{d_{wc}}{a_{wc}} \right)^2$ is the buckling coefficient for external connections

   $a_{wc}$ is the distance between web stiffeners. In the absence of web stiffeners, this value can be taken as infinity.
   $f_{y,c}$ is the yield stress of the column at elevated temperatures.

*Deformation capacity:*
According to, with reference to
   Figure 4.9 for dimensions:

$$\Delta_u = \frac{l_y t_{wc}^2}{2 b_{fc} t_{fc}} \sqrt{\frac{t_{wc} d_{wc}}{t_{fc}} c} \, \chi_F \tag{4.16}$$

## 4.3.8 Beam flange in compression(x)

*Stiffness and resistance:* as given in EN 1993-1-8 with modifications of material properties of steel at elevated temperatures.
   *Deformation capacity: infinite*

## 4.4 OTHER CONSIDERATIONS

The main aim of this book is to present a systematic approach to checking connection performance in fire, in particular, in the context of controlling progressive collapse. To do so, this book has collected a complete set of connection component behaviours for common types of open-section beam-column connections, for which the knowledge is relatively mature so that the entire process of checking connection performance in fire can be demonstrated.

   It should be appreciated that there are active developments to extend the component-based method to tubular connections (Weynand et al., 2003,

Jones and Wang, 2010, Park and Wang, 2012, Jaspart and Weynand, 2015, Jafarian and Wang, 2015). Furthermore, high-strength steel is being considered in building frame construction for which fire resistance is still being researched. Specialist steels such as stainless steel and fire-resistant steel can be used to improve connection performance in fire, in particular drastically enhancing connection deformation capacity (Chen and Wang, 2012, Elsawaf and Wang, 2012).

However, whilst the methodology of this book, adapted from the component-based method, can be used in the same way as described in this book for other types of connection using different materials, research is still in progress for these connections, and there are still many gaps of knowledge on these connections for them to be included in this book.

## REFERENCES

Beg, D., Zupančič, E. & Vayas, I. 2004. On the rotation capacity of moment connections. *Journal of Constructional Steel Research*, 60, 601–620.

Block, F. M. 2006. *Development of a component-based finite element for steel beam-to-column connections at elevated temperatures*. University of Sheffield, Sheffield, UK.

CEN 2004. EN 1994 Eurocode 4: Design of composite steel and concrete structures Part1-1: General rules and rules for buildings. British Standards Institute, London.

CEN 2005. EN 1993-1-8 Eurocode 3: Design of steel structures, Part 1–8: Design of joints. *British Standard*.

Chen, L. & Wang, Y. C. 2012. Efficient modelling of large deflection behaviour of restrained steel structures with realistic endplate beam/column connections in fire. *Engineering Structures*, 43, 194–209.

Elsawaf, S. & Wang, Y. C. 2012. Methods of improving the survival temperature in fire of steel beam connected to CFT column using reverse channel connection. *Engineering Structures*, 34, 132–146.

Hu, Y., Davison, B., Burgess, I. & Plank, R. 2009. Component modelling of flexible end-plate connections in fire. *International Journal of Steel Structures*, 9, 1–15.

Jafarian, M. & Wang, Y. 2015. Tying resistance of reverse channel connection to concrete filled square and rectangular tubular sections. *Engineering Structures*, 100, 17–30.

Jaspart, J. & Weynand, K. 2015. Design of hollow section joints using the component method. *Tubular Structures XV*, 403–410. doi: 10.1201/b18410-62.

Jones, M. H. & Wang, Y. C. 2010. Tying behaviour of fin-plate connection to concrete-filled rectangular steel tubular column — Development of a simplified calculation method. *Journal of Constructional Steel Research*, 66, 1–10.

Nethercot, D. A. 2006. Connection research and its impact on practice during the Dowling era. *Journal of Constructional Steel Research*, 62, 1165–1170.

Park, A. & Wang, Y. 2012. Development of component stiffness equations for bolted connections to RHS columns. *Journal of Constructional Steel Research*, 70, 137–152.

Sarraj, M. 2007. *The behaviour of steel fin plate connections in fire.* University of Sheffield.

Spyrou, S., Davison, J., Burgess, I. & Plank, R. 2002. Component studies for steelwork connections in fire. *5th International Conference on Stability and Ductility of Steel Structures*, Budapest, Hungary, 769–776.

Spyrou, S., Davison, J., Burgess, I. & Plank, R. 2004. Experimental and analytical investigation of the 'tension zone' components within a steel joint at elevated temperatures. *Journal of Constructional Steel Research*, 60, 867–896.

Taib, M. & Burgess, I. 2013. A component-based model for fin-plate connections in fire. *Journal of Structural Fire Engineering*, 4, 113–122.

Weynand, K., Jaspart, J. & Ly, L. 2003. Application of the component method to joints between hollow and open sections. *CIDECT Draft Final Report: 5BM.*

Yu, H., Burgess, I., Davison, J. & Plank, R. 2009a. Development of a yield-line model for endplate connections in fire. *Journal of Constructional Steel Research*, 65, 1279–1289.

Yu, H., Burgess, I., Davison, J. & Plank, R. 2009b. Tying capacity of web cleat connections in fire, Part 2: Development of component-based model. *Engineering Structures*, 31, 697–708.

Yu, H., Burgess, I. W., Davison, J. B. & Plank, R. J. 2009c. Experimental investigation of the behaviour of fin plate connections in fire. *Journal of Constructional Steel Research*, 65, 723–736.

# Chapter 5

# Application of the component-based method for beam–column connection design in fire

## 5.1 INTRODUCTION

Depending on the fire safety requirement of structures, different aspects of connection behaviour may be checked. In general, connections should be checked for their resistance against internal forces that are reaction forces to the connected beam. While it can be assumed that the shear force in the connected beam does not change in fire, as explained in Chapter 2 of this book, the internal axial force and bending moment in the connected beam vary as a function of time (or temperature of the connected beam). Therefore, connection checking depends on the stage of fire exposure to the connected beam, which in turn depends on the fire safety design requirement.

In a normal fire limit state design, the connected beam is assumed not to have any internal axial force and is in a stage of pure bending. This is represented by $\theta_{a, cr}$ in Figure 2.4. Furthermore, the connection is assumed to be pinned and, therefore, there is no need to check its bending resistance. The only required connection checking is shear resistance at the limiting temperature of the connected beam for bending resistance. These checks can be relatively easily performed following the same method as in ambient temperature design, but replacing the mechanical properties of connection components with those at elevated temperatures. Since the main focus of this book is about checking connection behaviour for structural robustness in fire, involving stages of structural behaviour in the catenary action stage, the examples in this chapter are to demonstrate these advanced checks. However, for completeness, Appendix B presents examples of common checks for shear and bending resistances of connections under normal fire limit state design.

Whilst there is no axial force in the connected beam at the limiting temperature of the beam under bending, the beam would develop compression force due to restrained thermal expansion before reaching the critical temperature for pure bending. Obtaining the exact value of this compression force involves prolonged calculations. Fortunately, as explained in Chapter 2, this is not necessary. For connections using endplates, there is no risk of connection failure during the compression phase of beam behaviour.

DOI: 10.1201/9781003134466-5

For connections that generate shear force in bolts when the connected beam is in compression (e.g. fin plate connection), there is a risk of premature failure of the connection before the beam reaches the intended limiting temperature under pure bending. However, this risk can be eliminated by ensuring that the failure mode of the connection is ductile, that is, any failure of the connection in the compression phase of the connected beam behaviour is bearing failure of the steel plate (e.g. fin plate or beam web) so that after reaching plastic failure of the ductile component, the compression force is released.

Therefore, it is the stage of the connected beam behaviour after reaching its limiting temperature that is the particular focus of this chapter. This checking may be required when dealing with the control of progressive collapse of the structure in fire: the connected beam may need to mobilise catenary action to resist the applied external load in fire to prevent progressive collapse. This is depicted in Figure 2.4 after the connected beam has exceeded its limiting temperature for pure bending at $\theta_{a,cr}$.

The purpose of checking the connection performance in the stage of catenary action of the connected beam is to calculate the maximum temperature of the beam until the fracture of a connection component, after which the risk of progressive collapse is high.

As shown in Figure 2.4, the behaviour of the beam in catenary action is different before and after reaching temperature at the peak catenary force, as indicated by $\theta_{peak}$ in Figure 2.4. Therefore, the following procedure is proposed for calculating the maximum beam temperature before reaching any fracture in the connection (survival temperature). For implementation, it is best using flowcharts to describe this procedure, as follows:

- Check the condition of the connection at $\theta_{peak}$ to decide whether the connection fails (which is defined as reaching the deformation capacity of the first connection component) before or after $\theta_{peak}$. Flowchart A (Figure 5.1) outlines the calculation steps, and the associated text presents detailed calculation equations.
- If the connection cannot enable the connected beam to reach $\theta_{peak}$, Flowchart B (Figure 5.2) and the associated detailed calculation equations should be followed.
- If the connection can enable the connected beam to develop pure catenary action beyond $\theta_{peak}$, Flowchart C (Figure 5.3) and the associated detailed calculation equations should be used.

All these detailed calculations start with a trial beam temperature. At this temperature, the connection is deformed to its maximum capacity. The associated internal forces in the connection and in the connected beam are calculated. At the survival temperature of the beam, the connection internal forces should be in equilibrium with the internal forces transmitted by the connected beam. The numerical examples in Section 5.3 are intended to demonstrate these detailed calculations.

**Flowchart A – Overall Method**

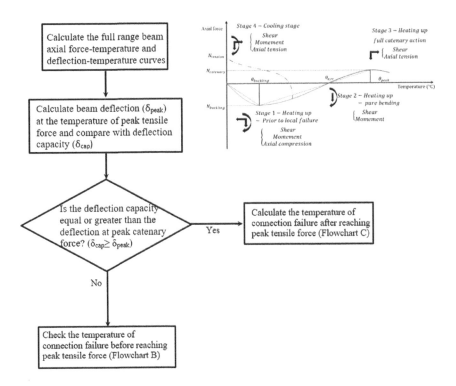

Calculate the full range beam axial force-temperature and deflection-temperature curves

Calculate beam deflection ($\delta_{peak}$) at the temperature of peak tensile force and compare with deflection capacity ($\delta_{cap}$)

Is the deflection capacity equal or greater than the deflection at peak catenary force? ($\delta_{cap} \geq \delta_{peak}$)

Yes

Calculate the temperature of connection failure after reaching peak tensile force (Flowchart C)

No

Check the temperature of connection failure before reaching peak tensile force (Flowchart B)

*Figure 5.1* Overall procedure.

As mentioned earlier in this chapter, it is also necessary to check the connection to resist the shear force in the connected beam. However, these calculations are easy to implement and are not demonstrated by numerical example in this chapter.

## 5.2 DETAILED IMPLEMENTATION OF CALCULATION PROCEDURES

### 5.2.1  Flowchart A: Detailed calculations to check connection at peak catenary action force

Beam deflection (demand) at the temperature of peak catenary force

$$\text{Temperature} = \theta_{peak} = (\theta_{a,\,cr} + 900)/2 \qquad (2.7)$$

**Flowchart B – check connection failure temperature before reaching peak catenary force ($\delta_{cap}$(supply) < $\delta_{peak}$(demand))**

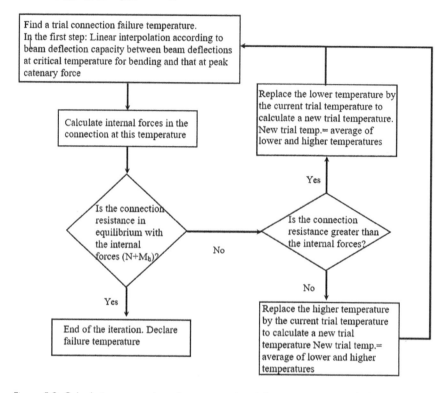

*Figure 5.2* Calculation procedure for connection failure temperature before reaching peak catenary force in the beam.

$M_E$ = maximum external applied moment in the beam under pinned supports at ends

$N_T$ = axial resistance of the cross-section

Deflection at the temperature $\delta_{peak} = M_E/F_T$    (2.10)

**Beam deflection capacity (supply)**

i. Deflection due to beam's thermal bowing without loading

Beam deflection due to thermal elongation $\delta_1 = (2L/\pi) \sqrt{(\alpha\Delta\theta + 0.5 (\alpha\Delta\theta)^2)}$ (Eqn. 2.6)

where

$\Delta\theta = \theta_{peak} - 20$

$\alpha$ = coefficient of thermal elongation of steel

$l$ = beam span

**Flowchart C – Check connection failure temperature after reaching peak catenary force**
($\delta_{cap}$ (supply) > $\delta_{peak}$ (demand))

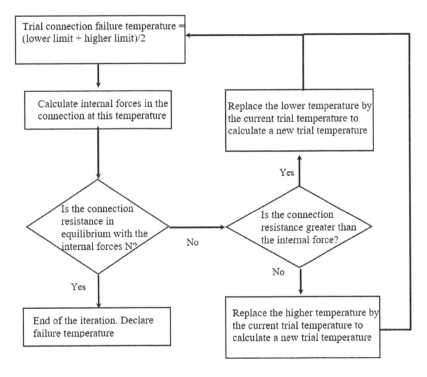

*Figure 5.3* Calculation procedure for connection failure temperature after reaching peak catenary force in the beam.

ii. Beam deflection due to rotation of the connection under loading
     Refer to Table 4.1 for the point of rotation of the connection and Figure 4.3 for dimensions.

For each series of active components, calculate the total displacement of all the active components corresponding to the maximum resistance of the weakest component. The weakest component reaches its deformation capacity while deformations of the other components are determined from their force–displacement curves.

For example, refer to Figure 5.4, for one series of spring components of an endplate connection (column flange in bending, bolt, endplate in bending) at a distance of h from the centre of rotation with Mode 3 failure (for other terms used in the following calculations, see Figure 4.3):

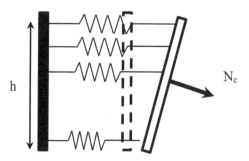

*Figure 5.4* State of loading in connection at peak catenary force.

Minimum resistance of the series (Mode 3 failure: bolts in tension)=F
Initial stiffness coefficient of the column flange in bending (EN 1993-1-8) $K_1=0.9\, l_{eff}\, t_f^3/m^3$
Horizontal deflection of the column flange in bending: $\Delta_1=F/(k_\theta EK_1)$
Initial stiffness coefficient of the endplate in bending (EN 1993-1-8) $K_2=0.9\, l_{eff}\, t_p^3/m^3$
Horizontal deflection of the endplate in bending: $\Delta_2=F/(k_\theta EK_2)$
Initial stiffness coefficient of the column web in tension (EN 1993-1-8) $K_3=0.7 b^{eff}\, t_f^3/d_c$
Horizontal deflection of the column web: $\Delta_3=F/(k_\theta EK_3)$
Deformation capacity of the bolts (equation 4.6) $\Delta_4=\varepsilon_{u,b}\,(L_b+2/n_{th})$
Total horizontal deformation of the series in tension: $\Delta=\Sigma\Delta_i=\Delta_1 + \Delta_2 + \Delta_3 + \Delta_4$
Total horizontal deformation capacity of the compression components: $\Delta_5$
Connection rotation of the series: $\varphi = (\Delta+\Delta_5)/h$

Perform the same calculations for all series of springs. The minimum rotation ($\varphi_{min}$) from all the series is that of the rotation capacity of the connection. The minimum rotation usually corresponds to the top role of bolts.

Maximum vertical deflection of the beam when reaching connection rotation capacity (equation 2.11): $\delta_2=L \tan (\varphi_{min})/4$
Deflection capacity of the beam: $\delta_{cap}=\delta_1+\delta_2$

## 5.2.2 Flowchart B – check connection failure temperature before reaching peak catenary force ($\delta_{cap}$(supply) $<\delta_{peak}$(demand))

i. Beam deflection at critical temperature for bending (Point 2 in Figure 2.4)
   Temperature=$\theta_{a,cr}$
   Deflection $\delta_{cr}=(2L/\pi) \sqrt{(\alpha\Delta\theta+0.5\,(\alpha\Delta\theta)^2)}$ (2.6)

ii. Linear interpolation to obtain a trial connection failure temperature

First step: linear interpolation for beam deflection capacity ($\delta_{cap}$) between beam deflections at the critical temperature for bending and the temperature of peak catenary force

$$\theta_{trial} = \left[\left(\delta_{cap} - \delta_{low}\right)/\left(\delta_{high} - \delta_{low}\right)\right]\left(\theta_{high} - \theta_{low}\right) + \theta_{low}$$

For step 1, output values from Flowchart A are used, and: $\theta_{low} = \theta_{a, cr}$, $\delta_{low} = \delta_{cr}$, $\theta_{high} = \theta_{peak}$, $\delta_{high} = \delta_{peak}$.

iii. Calculate connection component forces and the new beam deflection capacity $\delta_{trial}$ at $\theta_{trial}$

As for Flowchart A under beam deflection capacity, but replace $\theta_{peak}$ with $\theta_{trial}$.

iv. Internal forces at the trial temperature

At the trial temperature, the internal beam bending moments and catenary action force are as follows:

**Moment in the beam:** based on linear distribution between the critical temperature and the temperature at peak catenary force (equation 2.9a). The moment at critical temperature is the full plastic moment resistance $M_p$.

Therefore: $M_s = \left[-M_p / \left(\theta_{high} - \theta_{low}\right)\right]\left(\theta_{trial} - \theta_{low}\right) + M_p$

**Moment in the connection:** based on linear distribution between the critical temperature and the temperature at peak catenary force. The moment at critical temperature is $M_{h, cr} = M_E - M_s$.

Therefore: $M_h = \left[-M_{h, cr} / \left(\theta_{high} - \theta_{low}\right)\right]\left(\theta_{trial} - \theta_{low}\right) + M_{h, cr}$

Catenary action force (equation 2.7):

$$N = \left[M_E - \left(M_s + M_h\right)\right]/\delta_{trial}$$

where $M_E$ = maximum external applied moment in the beam under pinned supports at ends.

*note: it is not necessary to explicitly include the effect of axial force on bending moment. This is because the assumption of bending moment=0 at the peak catenary action force is to allow for the effect of axial load on bending moment. Therefore, the effect of axial load on bending moment is implicitly considered during linear interpolation.

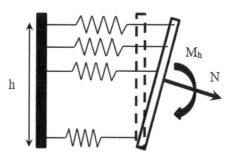

Figure 5.5 State of loading in connection under combined bending and catenary force in the beam.

v. Check connection resistance

In step (iii), limits of connection component resistance are calculated. These connection components' forces must be in equilibrium with the applied loads on the connection, which consist of the catenary action force N and the end moment $M_h$ from the beam, as shown in Figure 5.5.

For equilibrium check of the connection, take moment about the centre of rotation of the connection.

If the total anticlockwise moment of all spring elements is equal to the total clockwise moment from the catenary action force (N) and the beam end moment ($M_h$), then the connection is at the critical stage of reaching failure. The trial temperature is the failure temperature of the connection.

If the total anticlockwise moment of all spring elements is (a) less or (b) higher than the total clockwise moment from the catenary action force (N) and the beam end moment ($M_h$), then the connection failure temperature is, respectively, lower or higher than the current trial temperature and iteration is needed. For the next iteration, replace the higher limit temperature with the current trial temperature for case (a) and replace the lower limit temperature by the current trial temperature for case (b). The new trial temperature is the average value of the new lower and higher limits. Then continue from step (iii).

### 5.2.3 Flowchart C – Check connection failure temperature after reaching peak catenary force ($\delta_{cap}$(supply) > $\delta_{peak}$(demand))

i. Trial connection failure temperature

$$\theta_{trial} = \left(\theta_{low} + \theta_{high}\right)/2. \text{ For step 1, } \theta_{low} = \theta_{peak}, \theta_{high} = 900°C$$

*note: interpolation based on deflections as in Flowchart B is not possible in this case because the deflection at 900°C is infinite by definition.

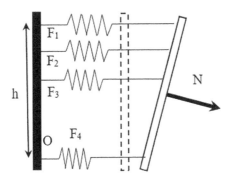

*Figure 5.6* State of loading in connection under pure catenary action in the beam.

  ii. Calculate connection component forces and new beam deflection capacity $\delta_{\text{trial}}$ at $\theta_{\text{trial}}$

      As for Flowchart A under beam deflection capacity, replace $\theta_{\text{peak}}$ with $\theta_{\text{trial}}$.

 iii. Internal forces at the trial temperature

$$M_s = M_b = 0$$
$$N = M_E / \delta_{\text{trail}}$$

      where $M_E$ = maximum external applied moment in the beam under pinned supports at ends

  iv. Check connection resistance

      In step (ii), limits of connection component resistance are calculated. These connection components' forces must be in equilibrium with the applied load on the connection, which consists of only the catenary action force N, as shown in Figure 5.6.

For equilibrium check of the connection, take moment about the centre of rotation of the connection.

If the total anticlockwise moment of all spring elements is equal to the total clockwise moment from the catenary action force (N), then the connection is at the critical stage of reaching failure. The trial temperature is the failure temperature of the connection.

If the total anticlockwise moment of all spring elements is (a) less or (b) higher than the total clockwise moment from the catenary action force (N), then the connection failure temperature is, respectively, lower or higher than the current trial temperature and iteration is needed. For the next iteration, replace the higher limit temperature with the current trial temperature for case (a), and replace the lower limit temperature with the current trial temperature for case (b). The new trial temperature is the average value of the new lower and higher limits. Then continue from step (i).

## 5.3 NUMERICAL EXAMPLES

Although it is possible for all connections to allow the connected beam to develop some degree of catenary action in the connected beam, only connections with high rotation/deformation capacity would realistically enable the connected beam to develop sufficient catenary action to survive temperatures much higher (e.g. >20% of the limiting temperature) than the beam's limiting temperature for bending. Under this circumstance, it is not good practice to propose connections with low rotation/deformation capacity (e.g. fin plate connection) to resist progressive collapse. Instead, connections with high rotation/deformation capacity (such as end-plate connections) should be used. Therefore, even though the procedure for checking connection is the same for different types of connection, the numerical examples of this section are for extended endplate connections only.

As explained in Chapter 3, connection components usually have lower temperatures than the temperature of the connected beam. For simplicity, in the examples of this chapter, the connection is assumed to have the same temperature as the connected beam.

Appendix B provides mechanical property data of different types of steel at elevated temperatures, extracted from EN 1993-1-2 (CEN, 2005) from which the data in the various examples are obtained.

### 5.3.1  Basic data

The connection used in this worked example is taken from the worked example in SCI P398 (Brown et al., 2013). The connection is between a primary beam and a column of an office building shown in Figure 5.7a.

Connection dimensions and materials are as follows.

| Bolts (M24 Grade 8.8) | Column flange | Endplate |
|---|---|---|
| Diameter of bolt shank $(d) = 24$ mm | Spacing $(w) = 100$ mm | End distance $(e_x) = 50$ mm |
| Diameter of hole $(d_0) = 26$ mm | Edge distance $(e_c) = 0.5 \times (258.8 - 100) = 79.4$ mm | Spacing $(w) = 100$ mm |
| Shear area $(A_s) = 353$ mm$^2$ | Spacing between rows 1 and 2 $(p_{1-2}) = 100$ mm | Edge distance $(e_p) = 75$ mm |
| Diameter of washer $d_w = 41.6$ mm | Spacing between rows 2 and 3 $(p_{2-3}) = 90$ mm | Distance between row 1 and beam flange $(x) = 40$ mm |
| | | Spacing between rows 1 and 2 $(p_{1-2}) = 100$ mm |
| | | Spacing between rows 2 and 3 $(p_{2-3}) = 90$ mm |

Yield stress of steel for the beam: $f_{y,b} = 275$ N/mm$^2$
Yield stress of steel for the column and endplate: $f_{y,c} = 265$ N/mm$^2$
Yield stress of bolt: $f_{yb} = 640$ N/mm$^2$
Ultimate tensile stress of bolt: $f_{ub} = 800$ N/mm$^2$

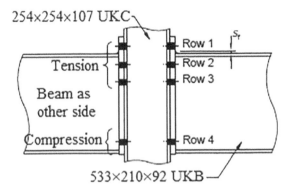

*Figure 5.7* Structural floor layout and connection details. (a) Plan layout. (b) Connection dimensions.

Applied load (centre point load) on the primary beam: $P = 324.5$ kN

Maximum shear force in the primary beam: $V_{Ed,fi} = R_{fi}/2 = 324.5/2 = 162.25$ kN

Total maximum bending moment in the beam under pin supports at the end:

$$M_E = 324.5 \times 6/4 = 486.75 \text{ kN} \cdot \text{m}$$

The SCI guide for ambient temperature design gives a connection bending moment resistance of about 50% of the sagging moment resistance of the beam. Assume this ratio is maintained at the fire limit state for pure bending. Therefore,

Hogging moment $M_h = 1/3 \times 324.5 = 108.17$ kN·m

Sagging moment $M_s = 2/3 \times 649 = 216.33$ kN·m

### 5.3.2 Beam axial force – temperature and maximum deflection–temperature curves

The same connection was used as Example 2.2 to evaluate the beam axial force–temperature and beam maximum deflection–temperature curves. These curves are repeated as in Figure 5.8a and b.

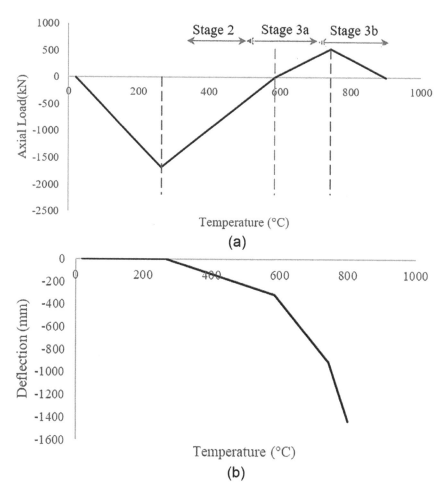

*Figure 5.8* Load and deflection – temperature curves. (a) Axial force – temperature. (b) deflection – temperature.

### 5.3.3 Checking the connection at peak catenary action force (Flowchart A)

Critical temperature: 584.7°C (Chapter 2)
Temperature at peak catenary action force: $=(584.7 + 900)/2 = 742.3°C$
Yield and ultimate stress reduction factor for bolt: $k_{b,\theta} = 0.086$
Yield stress reduction factor for steel: $k_{y,\theta} = 0.1792$
Young's modulus reduction factor for steel: $k_{E,\theta} = 0.113$
Detailed calculation results for connection components:

**Row 1**

Resistance

## Column flange in bending

### Mode 1 (four plastic hinges)

$F_{T,1,Rd} = (8n - 2e_w) M_{pl,fi,Rd}/(2 mn - ew (m+n))$

$M_{pl,fi,Rd} = 0.25 k_y \Sigma l_{eff,1} t_f^2 fy/\gamma_{fi} = 0.25 \times 0.1792 \times 210 \times 20.5^2 \times 265/1.0 = 1047.7$ kN·mm

$e_w = d_w/4 = 39.55/4 = 9.9$ mm

$F_{T,1,Rd} = ((8 \times 41.8 - 2 \times 9.9) \times 1047.7)/(2 \times 33.4 \times 41.8 - 9.9 \times (33.4+41.8)) = 160.69$ kN

### Mode 2 (two plastic hinges and bolt failure)

$F_{T,2,Rd} = (2 M_{pl,fi,Rd} + n \Sigma_{Ft,b,Rd})/(m + n)$

$M_{pl,fi,Rd} = 0.25 ky \Sigma l_{eff,2} t_f^2 f_y/\gamma_{fi} == 0.25 \times 0.1792 \times 233 \times 20.5^2 \times 265/1.0 = 1163.014$ kN·mm

$F_{t,b,Rd} = (k_2 k_p b f_{ub} A_s)/\gamma_{fi} = 0.9 \times 0.086 \times 800 \times 353/1.0 = 21.87$ kN

$F_{T,2,Rd} = (2 \times 1163.014 + 41.48 \times (2 \times 21.87))/(33.4+41.8) = 55.05$ kN

### Mode 3 (bolt failure)

$F_{T,3,Rd} = \Sigma F_{t,b,Rd} = 2 \times (0.9 \times 0.086 \times 800 \times 353/1.0) = 43.73$ kN

**Resistance = Min (Mode 1, Mode 2, Mode 3) = 43.73 kN**

Column flange-Stiffness (EN1993-1-8 formula) $(k_E E) (0.9 l_{eff} t_f^3/m^3)$

$= 0.113 \times 205000 \times 0.9 \times 210 \times 20.5^3/33.4^3 = 1.01 \times 10^6$ N/mm

## Column web in transverse tension

$F_{t,wc,Rd} = \omega b_{eff,t,wc} t_{wc} k_y f_y/\gamma_{fi} = 1.0 \times 233 \times 12.8 \times 0.1792 \times 265/1000 = 141.63$ kN

Column web-Stiffness (EN1993-1-8 formula) $= (k_E E) (0.7 b_{eff,t,wc} t_{wc} / d_c) =$

$0.113 \times 205000 \times 0.7 \times 233 \times 12.8/200.3 = 0.242 \times 10^6$ N/mm

Endplate in bending

### Mode 1 (four plastic hinges)

$F_{T,1,Rd} = (8n - 2e_w) M_{pl,fi,Rd} / (2mn - ew (m+n))$

$M_{pl,fi,Rd} = 0.25 k_y \Sigma l_{eff,1} t_p^2 f_y/\gamma_{fi} = 0.25 \times 0.1792 \times 125 \times 25^2 \times 265/1.0 = 927.5$ kN·mm

$e_w = d_w/4 = 39.55/4 = 9.9$ mm

$F_{T,1,Rd} = ((8 \times 38 - 2 \times 9.9) \times 927.5)/(2 \times 30.4 \times 38 - 9.9 \times (30.4+38)) = 161.39$ kN

Row 2

**Mode 2 (two plastic hinges and bolt failure)**

$F_{T,2,Rd} = (2 M_{pl,fi,Rd} + n \Sigma F_{t,b,Rd})/(m + n)$

$M_{pl,fi,Rd} = 0.25 k_y \Sigma l_{eff,2} t_f^2 f_y / \gamma_{fi} = 0.25 \times 0.1792 \times 125 \times 25^2 \times 265/1.0 = 927.5$ kN·mm

$F_{t,b,Rd} = (k_2 k_{y,b} f_{ub} A_s) / \gamma_{fi} = 0.9 \times 0.086 \times 800 \times 353/1.0 = 21.87$ kN

$F_{T,2,Rd} = (2 \times 927.5 + 38 \times (2 \times 21.87))/(30.4+38) = 51.44$ kN

**Mode 3 = 94.54 kN (bolt failure)**

$F_{T,3,Rd} = \Sigma F_{t,b,Rd} = 2 \times (0.9 \times 0.086 \times 800 \times 353/1.0) = 43.73$ kN

**Resistance = Min (Mode 1, Mode 2, Mode 3) = 43.73 kN**

Endplate-Stiffness (EN1993-1-8 formula) = $(k_E E) (0.9 \, l_{eff} \, t_p^3/m^3) =$
$0.113 \times 205000 \times 0.9 \times 125 \times 25^3/30.4^3 = 1.45 \times 10^6$ N/mm

Row 2

**Column flange in bending**
**Mode 1 (four plastic hinges)**

$F_{T,1,Rd} = (8n - 2e_w) M_{pl,fi,Rd} / (2 mn - e_w (m + n))$

$M_{pl,fi,Rd} = 0.25 k_y \Sigma l_{eff,1} t_f^2 f_y/\gamma_{fi} = 0.25 \times 0.1792 \times 210 \times 20.5^2 \times 265/1.0 = 1047.7$ kN·mm

$e_w = d_w/4 = 39.55/4 = 9.9$mm

$F_{T,1,Rd} = ((8.41.8 - 2 \times 9.9) \times 1047.7)/(2 \times 33.4 \times 41.8 - 9.9 \times (33.4 + 41.8)) = 160.96$ kN

**Mode 2 (two plastic hinges and bolt failure)**

$F_{T,2,Rd} = (2 M_{pl,fi,Rd} + n \Sigma F_{t,b,Rd})/(m + n)$

$M_{pl,fi,Rd} = 0.25 k_y \Sigma l_{eff,2} t_f^2 f_y/\gamma_{fi} = 0.25 \times 0.1792 \times 233 \times 20.5^2 \times 265/1.0 = 1163.014$ kN·mm

$F_{t,b,Rd} = (k_2 k_{y,b} f_{ub} A_s)/\gamma_{fi} = 0.9 \times 0.086 \times 800 \times 353/1.0 = 21.87$ kN

$F_{T,2,Rd} = (2 \times 1163.014 + 41.48 \times (2 \times 21.87))/(33.4 + 41.8) = 55.05$ kN

**Mode 3 (bolt failure)**

$F_{T,3,Rd} = \Sigma F_{t,b,Rd} = 2 \times (0.9 \times 0.086 \times 800 \times 353/1.0) = 43.73$ kN

**Resistance = Min (Mode 1, Mode 2, Mode 3) =43.73 kN**

Column flange-stiffness (EN1993-1-8 formula) = $(k_E E) (0.9 \, l_{eff} \, t_f^3/m^3) =$
$0.113 \times 205000 \times 0.9 \times ((100+90)/2) \times 20.5^3/33.4^3 = 0.46 \times 10^6$ N/mm

## Column web in transverse tension

$F_{t,wc,Rd} = \omega\, b_{eff,t,wc}\, t_{wc}\, k_y\, f_y/\gamma_{fi} = 1.0 \times 233 \times 12.8 \times 0.1792 \times 265/1000 = 141.63$ kN

Column web-stiffness $= (k_E\, E)\,(0.7\, b_{eff,t,wc}\, t_{wc}/d_c)\; 0.113 \times 205000 \times 0.7 \times ((100+90)/2) \times$
$12.8/200.3 = 0.099 \times 10^6$ N/mm

## Endplate in bending

### Mode I (four plastic hinges)

$F_{T,1,Rd} = (8n - 2e_w)\, M_{pl,fi,Rd} / (2mn - e_w\,(m + n))$

$M_{pl,fi,Rd} = 0.25\, k_y\, \Sigma l_{eff,1}\, t_p^2\, f_y / \gamma_{fi} = 0.25 \times 0.1792 \times 243 \times 25^2 \times 265/1.0 = 1803.06$ kN·mm

$e_w = d_w / 4 = 39.55/4 = 9.9$ mm

$F_{T,1,Rd} = ((8 \times 48.3 - 2 \times 9.9) \times 1803.06)/(2 \times 38.6 \times 48.3 - 9.9 \times (38.6 + 48.3)) = 230.44$ kN

### Mode 2 (two plastic hinges and bolt failure)

$F_{T,2,Rd} = (2\, M_{pl,fi,Rd} + n\, \Sigma F_{t,b,Rd})/(m + n)$

$M_{pl,fi,Rd} = 0.25\, k_y\, \Sigma l_{eff,2}\, t_p^2\, f_y/\gamma_{fi} = 0.25 \times 0.1792 \times 290 \times 25^2 \times 265/1.0 = 2151.8$ kN·mm

$F_{t,b,Rd} = (k_2\, k_{y,b}\, f_{ub}\, A_s)/\gamma_{fi} = 0.9 \times 0.086 \times 800 \times 353/1.0 = 21.87$ kN

$F_{T,2,Rd} = (2 \times 2151.8 + 48.3 \times (2 \times 21.87))/(38.6 + 48.3) = 73.8$ kN

### Mode 3 (bolt failure)

$F_{T,3,Rd} = \Sigma F_{t,b,Rd} = 2 \times (0.9 \times 0.086 \times 800 \times 353/1.0) = 43.73$ kN

### Resistance = Min (Mode I, Mode 2, Mode 3) = 43.73 kN

Endplate-Stiffness (EN1993-1-8 formula) $= (k_E\, E)\,(0.9\, l_{eff}\, t_p^3/m^3) =$
$0.113 \times 205000 \times 0.9 \times 243 \times 25^3/38.6^3 = 1.37 \times 10^6$ N/mm

## Beam web in tension

$F_{t,wb,Rd} = \omega\, b_{eff,t,wc}\, t_{wb}\, k_y\, f_y/\gamma_{fi} = 1.0 \times 243 \times 10.1 \times 0.1792 \times 275/1000 = 120.95$ kN

## Row I and 2 combined – column flange in bending

### Column flange in bending

### Mode I (four plastic hinges)

$F_{T,1,Rd} = (8n - 2e_w)\, M_{pl,fi,Rd}/(2mn - ew\,(m + n))$

$M_{pl,fi,Rd} = 0.25\, k_y\, \Sigma l_{eff,1}\, t_f^2\, f_y/\gamma_{fi} = 0.25 \times 0.1792 \times 332 \times 20.5^2 \times 265/1.0 = 1656.42$ kN·mm

$e_w = d_w / 4 = 39.55/4 = 9.9$ mm

$F_{T,1,Rd} = ((8 \times 41.8 - 2 \times 9.9) \times 1656.42)/(2 \times 33.4 \times 41.8 - 9.9 \times (33.4 + 41.8)) = 255.71$ kN

Row 3

**Mode 2 (two plastic hinges and bolt failure)**

$F_{T,2,Rd} = (2\,M_{pl,fi,Rd} + n\,\Sigma F_{t,b,Rd})/(m+n)$

$M_{pl,fi,Rd} = 0.25\,k_y\,\Sigma l_{eff,2}\,t_f^2\,f_y/\gamma_{fi} = 0.25 \times 0.1792 \times 332 \times 20.5^2 \times 265/1.0 = 1656.42$ kN·mm

$F_{t,b,Rd} = (k_2\,k_{y,b}\,f_{ub}\,A_s)/\gamma_{fi} = 0.9 \times 0.086 \times 800 \times 353/1.0 = 21.87$ kN

$F_{T,2,Rd} = (2 \times 1656.42 + 41.8 + (4 \times 21.87))/(33.4 + 41.8) = 92.68$ kN

**Mode 3 (bolt failure)**

$F_{T,3,Rd} = \Sigma F_{t,b,Rd} = 4 \times (0.9 \times 0.086 \times 800 \times 353/1.0) = 87.46$ kN

**Resistance = Min (Mode 1, Mode 2, Mode 3) = 87.46 kN**

Row 1 and 2 combined – Column web in transverse tension

$F_{t,wc,Rd} = \omega\,b_{eff,t,wc}\,t_{wc}\,k_y\,f_y/\gamma_{fi} = 1.0 \times 332 \times 12.8 \times 0.1792 \times 265/1000 = 201.78$ kN

Therefore, the resistance of the bolt row 2 on the column side is = **Min (87.46, 201.78) – 43.73 = 43.73 kN**

**Column flange in bending**

**Mode 1 (four plastic hinges)**

$F_{T,1,Rd} = (8n - 2e_w)\,M_{pl,fi,Rd}/(2mn - ew\,(m+n))$

$M_{pl,fi,Rd} = 0.25\,k_y\,\Sigma l_{eff,1}\,t_f^2\,f_y/\gamma_{fi} = 0.25 \times 0.1792 \times 210 \times 20.5^2 \times 265/1.0 = 1047.7$ kN·mm

$e_w = d_w/4 = 39.55/4 = 9.9$ mm

$F_{T,1,Rd} = ((8 \times 41.8 - 2 \times 9.9) \times 1047.7)/(2 \times 33.4 \times 41.8 - 9.9 \times (33.4 + 41.8)) = 160.96$ kN

**Mode 2 (two plastic hinges and bolt failure)**

$F_{T,2,Rd} = (2\,M_{pl,fi,Rd} + n\,\Sigma F_{t,b,Rd})/(m+n)$

$M_{pl,fi,Rd} = 0.25\,k_y\,\Sigma l_{eff,2}\,t_f^2\,f_y/\gamma_{fi} = 0.25 \times 0.1792 \times 233 \times 20.5^2 \times 265/1.0 = 1162.49$ kN·mm

$F_{t,b,Rd} = (k_2\,k_{y,b}\,f_{ub}\,A_s)/\gamma_{fi} = 0.9 \times 0.086 \times 800 \times 353/1.0 = 21.87$ kN

$F_{T,2,Rd} = (2 \times 1162.49 + 41.48 + (2 \times 21.87))/(33.4 + 41.8) = 55.04$ kN

**Mode 3 (bolt failure)**

$F_{T,3,Rd} = \Sigma F_{t,b,Rd} = 2 \times (0.9 \times 0.086 \times 800 \times 353/1.0) = 43.73$ kN

**Resistance = Min (Mode 1, Mode 2, Mode 3) = 43.73 kN**

Column flange-Stiffness (ENI 993-1-8 formula) $= 0.113 \times 205000 \times 0.9 \times 161.43 \times 20.5^3/33.4^3 = 0.78 \times 10^6$ N/mm

Row 3

## Column web in transverse tension

$F_{t,wc,Rd} = \omega b_{eff,t,wc} t_{wc} k_y f_y / \gamma_{fi} = 1.0 \times 233 \times 12.8 \times 0.1792 \times 265/1000 = 141.63$ kN

Column web-stiffness (EN1993-1-8 formula) $= (k_E E) (0.7\ b_{eff,t,wc}\ t_{wc} / d_c) =$

$0.113 \times 205000 \times 0.7 \times 161.43 \times 12.8/200.3 = 0.167 \times 10^6$ N/mm

## Endplate in bending
### Mode 1 (four plastic hinges)

$F_{T,1,Rd} = (8n - 2e_w) M_{pl,fi,Rd} / (2mn - ew (m+n))$

$M_{pl,fi,Rd} = 0.25\ k_y\ \Sigma l_{eff,1}\ t_p^2\ f_y/\gamma_{fi} = 0.25 \times 0.1792 \times 243 \times 25^2 \times 265/1.0 = 1803.06$ kN·mm

$e_w = d_w / 4 = 39.55/4 = 9.9$ mm

$F_{T,1,Rd} = ((8 \times 48.3 - 2 \times 9.9) \times 1803.06)/(2 \times 38.6 \times 48.3 - 9.9 \times (38.6 + 48.3)) = 230.44$ kN

### Mode 2 = (two plastic hinges and bolt failure)

$F_{T,2,Rd} = (2\ M_{pl,fi,Rd} + n\ \Sigma_{t,b,Rd})/(m+n)$

$M_{pl,fi,Rd} = 0.25\ k_y\ \Sigma l_{eff,2}\ t_p^2\ f_y/\gamma_{fi} = 0.25 \times 0.1792 \times 248 \times 25^2 \times 265/1.0 = 1840.16$ kN·mm

$F_{t,b,Rd} = (k_2\ k_{y,b}\ f_{ub}\ A_s)/\gamma_{fi} = 0.9 \times 0.086 \times 800 \times 353/1.0 = 21.87$ kN

$F_{T,2,Rd} = (2 \times 1840.16 \times 10^3 + 48.3 \times (2 \times 21.87))/(38.6 + 48.3) = 66.66$ kN

### Mode 3 (bolt failure)

$\Sigma F_{t,b,Rd} = 2 \times (0.9 \times 0.086 \times 800 \times 353/1.0) = 43.73$ kN

## Resistance = Min (Mode 1, Mode 2, Mode 3) = 43.73 kN

Endplate-Stiffness (EN1993-1-8 formula) $= (k_E E) (0.9\ l_{eff}\ t_p^3/m^3) =$

$0.113 \times 205000 \times 0.9 \times 243 \times 25^3/38.6^3 = 1.377 \times 10^6$ N/mm

## Beam web in tension

$F_{t,wb,Rd} = \omega b_{eff,t,wc} t_{wb} k_y f_y/\gamma_{fi} = 1.0 \times 243 \times 10.1 \times 0.1792 \times 275/1000 = 120.95$ kN

## Row 1, 2 and 3 combined – column flange in bending
### Column flange in bending
### Mode 1 (four plastic hinges)

$F_{T,1,Rd} = (8n - 2e_w) M_{pl,fi,Rd} / (2mn - ew (m+n))$

$M_{pl,fi,Rd} = 0.25\ k_y\ \Sigma l_{eff,1}\ t_f^2\ f_y/\gamma_{fi} = 0.25 \times 0.1792 \times 422 \times 20.5^2 \times 265/1.0 = 2105.44$ kN·mm

$e_w = d_w / 4 = 39.55/4 = 9.9$ mm

$F_{T,1,Rd} = ((8 \times 41.8 - 2 \times 9.9) \times 2105.44)/(2 \times 33.4 \times 41.8 - 9.9 \times (33.4 + 41.8)) = 323.46$ kN

**Mode 2 (two plastic hinges and bolt failure)**

$F_{T,2,Rd} = (2\,M_{pl,fi,Rd} + n\,\Sigma F_{t,b,Rd})/(m+n)$

$M_{pl,fi,Rd} = 0.25\,k_y\,\Sigma l_{eff,2}\,t_f^2\,f_y/\gamma f_i = 0.25 \times 0.1792 \times 422 \times 20.5^2 \times 265/1.0 = 2105.44$ kN·mm

$F_{t,b,Rd} = (k_2\,k_{y,b}\,f_{ub}^A\,A_s)/\gamma f_i = 0.9 \times 0.086 \times 800 \times 353/1.0 = 21.87$ kN

$F_{T,2,Rd} = (2 \times 2105.44 + 41.8 \times (6 \times 21.87))/(33.4 + 41.8) = 128.69$ kN

**Mode 3 (bolt failure)**

$F_{T,3,Rd} = \Sigma F_{t,b,Rd} = 6 \times (0.9 \times 0.086 \times 800 \times 353/1.0) = 131.19$ kN

**Resistance = Min (Mode 1, Mode 2, Mode 3) = 128.69 kN**

Row 1, 2 and 3 combined – Column web in transverse tension

$F_{t,wc,Rd} = \omega\,b_{eff,t,wc}\,t_{wc}\,k_y\,f_y/\gamma f_i = 1.0 \times 422 \times 12.8 \times 0.1792 \times 265/1000 = 256.51$ kN

**Therefore, the resistance of the bolt row 3 on the column side is = Min (128.69, 256.51) – 87.46 = 41.22 kN**

**Row 2 and 3 combined – column flange in bending**

**Mode 1 (four plastic hinges)**

$F_{T,1,Rd} = (8n - 2e_w)\,M_{pl,fi,Rd}/(2mn - ew\,(m+n))$

$M_{pl,fi,Rd} = 0.25\,k_y\,\Sigma l_{eff,1}\,t_f^2\,f_y/\gamma f_i = 0.25 \times 0.1792 \times 323 \times 20.5^2 \times 265/1.0 = 1611.51$ kN·mm

$e_w = d_w/4 = 39.55/4 = 9.9$ mm

$F_{T,1,Rd} = ((8 \times 41.8 - 2 \times 9.9) \times 1611.51)/(2 \times 33.4 \times 41.8 - 9.9 \times (33.4 + 41.8)) = 247.58$ kN

**Mode 2 (two plastic hinges and bolt failure)**

$F_{T,2,Rd} = (2\,M_{pl,fi,Rd} + n\,\Sigma F_{t,b,Rd})/(m+n)$

$M_{pl,fi,Rd} = 0.25\,k_y\,\Sigma l_{eff,2}\,t_f^2\,f_y/\gamma f_i = 0.25 \times 0.1792 \times 323 \times 20.5^2 \times 265/1.0 = 1611.51$ kN·mm

$F_{t,b,Rd} = (k_2\,k_{y,b}\,f_{ub}^A\,A_s)/\gamma f_i = 0.9 \times 0.086 \times 800 \times 353/1.0 = 21.87$ kN

$F_{T,2,Rd} = (2 \times 1611.51 + 41.8 \times (4 \times 21.87))/(33.4 + 41.8) = 91.49$ kN

**Mode 3 (bolt failure)**

$F_{T,3,Rd} = \Sigma F_{t,b,Rd} = 4 \times (0.9 \times 0.086 \times 800 \times 353/1.0) = 87.46$ kN

**Resistance = Min (Mode 1, Mode 2, Mode 3) = 87.46 kN**

Row 2 and 3 combined – column web in transverse tension

$F_{t,wc,Rd} = \omega_{beff,t,wc}\,t_{wc}\,k_y\,f_y/\gamma f_i = 1.0 \times 323 \times 12.8 \times 0.1792 \times 265/1000 = 196.33$ kN

**Therefore, bolt row 3 resistance on the column side = Min (87.46, 196.33) − 43.73 = 43.73 kN**

**Row 2 and 3 combined – beam side endplate in bending**

**Mode 1 (four plastic hinges)**

$F_{T,1,Rd} = (8n - 2e_w) M_{pl,fi,Rd}/(2mn - ew (m + n))$

$M_{pl,fi,Rd} = 0.25 \, k_y \, \Sigma l_{eff,1} \, t_f^2 \, f_y/\gamma f_i = 0.25 \times 0.1792 \times 379 \times 25^2 \times 265/1.0 = 2812.18$ kN·mm

$e_w = d_w/4 = 39.55/4 = 9.9$ mm

$F_{T,1,Rd} = ((8 \times 48.3 - 2 \times 9.9) \times 2812.18)/(2 \times 38.6 \times 48.3 - 9.9 \times (38.6 + 48.3)) = 359.41$ kN

**Mode 2 (two plastic hinges and bolt failure)**

$F_{T,2,Rd} = (2 \, M_{pl,fi,Rd} + n \, \Sigma_{Ft,b,Rd})/(m + n)$

$M_{pl,fi,Rd} = 0.25 \, k_y \, \Sigma l_{eff,2} \, t_f^2 \, f_y/\gamma f_i = 0.25 \times 0.1792 \times 379 \times 25^2 \times 265/1.0 = 2812.18$ kN·mm

$F_{t,b,Rd} = (k_2 \, k_{y,b} \, f_{ub} \, A_s)/\gamma f_i = 0.9 \times 0.086 \times 800 \times 353/1.0 = 21.87$ kN

$F_{T,2,Rd} = (2 \times 2812.18 + 48.3 \times (4 \times 21.87))/(38.6 + 48.3) = 113.34$ kN

**Mode 3 (bolt failure)**

$F_{T,3,Rd} = \Sigma F_{t,b,Rd} = 4 \times (0.9 \times 0.086 \times 800 \times 353/1.0) = 87.46$ kN

**Resistance = Min (Mode 1, Mode 2, Mode 3) = 87.46 kN**
**Therefore, the resistance of the bolt row 3 on the column side is = 87.46 − 43.73 = 43.73 kN**

**Compression zone**

Column web in transverse compression = $F_{c,wc,Rd} = \omega_{kwc} \, b_{eff,c,wc} \, t_{wc} \, k_y \, f_y \, /\gamma f_i \le \omega \, k_{wc} \, \rho \, b_{eff,c,wc} \, t_{wc} \, k_y \, f_y \, /$ Compression zone

$\gamma f_i \, \sqrt{k_y} \, / \, k_E = 1.25; \sqrt{\lambda} = 0.685; \rho = 1.0; b_{eff,c,wc} = 248$ mm; $k_{wc} = 1.0$

$F_{c,wc,Rd} = 1.0 \times 1.0 \times 248 \times 12.8 \times 0.1792 \times 265/1.0 = 150.75$ kN

Column web in transverse compression - stiffness (equation 4.4 of Chapter 4) = $0.95 \, k_E \, E \, (b_{fc} \, t_{wc}^2$

$t_{fc}^3 \, / \, (b_{eff} \, d_{wc}))^{0.25} = 0.95 \times 0.113 \times 205000 \times (258.8 \times 12.8^2 \times 20.5^3/248 \times 200.3)^{0.25} = 0.204 \times 10^6$ N/mm

**Beam flange and web in compression**

$F_{c,fb,Rd,fi} = M_{c,Rd,fi} \, / \, (h - t_{fb}) = 649 \times 0.1792/(533.1 - 15.6) = 224.74$ kN

Initial stiffness of beam flange and column web = Infinity

Initial stiffness of bolts in tension = $2 \times 0.086 \times 205000 \times 353/65.5 = 0.19 \times 10^6$ N/mm

A summary of the calculation results is presented in Table 5.1.

The governing failure mode, in this case, is failure Mode 3 (i.e. bolt fracture).

**Maximum beam deflection and connection component forces:**
Row 1

Bolt deflection at failure (equation 4.6) = $\varepsilon_{u,b}$ $(L_b+2/n_{th})$ = $0.20 \times 52.5$ = 10.5 mm

Column flange deflection (equation 4.1) = $F_y/K_{cf}$ = $43.73 \times 10^3/1.01 \times 10^6$ = 0.043 mm

Column web deflection (Section 4.3.2) = $F_y/K_{cw}$ = $43.73 \times 10^3/0.242 \times 10^6$ = 0.181 mm

Endplate deflection (equation 4.1) = $F_y/K_{cf}$ = $43.73 \times 10^3/1.45 \times 10^6$ = 0.0305 mm

Deformation of compression components: negligibly small

Total row 1 deformation: 10.75 mm

Rotation of the connection due to row 1 deformations: $10.75/573 = 0.0188$

Row 2

Bolt deflection at failure (equation 4.6) = $\varepsilon_{u,b}$ $(L_b+2/n_{th})$ = $0.20 \times 52.5$ = 10.5 mm

Column flange deflection (equation 4.1) = $F_y/K_{cf}$ = $43.73 \times 10^3/0.458 \times 10^6 = 0.095$ mm

Column web deflection (Section 4.3.2) = $F_y/K_{cw}$ = $43.73 \times 10^3/0.0985 \times 10^6$ = 0.444 mm

Endplate deflection (equation 4.1) = $F_y/K_{cf}$ = $43.73 \times 10^3/1.37 \times 10^6$ = 0.032 mm

Total row 2 deformation: 11.07

Rotation of the connection due to row 2 deformations: $11.07/473 = 0.023$

Row 3

Bolt deflection at failure (equation 4.6) at 41.22 kN (0.94 of the ultimate capacity) = $\varepsilon_{u,b}$ $(L_b+2/n_{th})=0.20 \times 52.5 \times 0.94=10.5$ mm

Column flange deflection (equation 4.1) = $F_y/K_{cf}$ = $41.22 \times 10^3/0.779 \times 10^6 = 0.053$ mm

Column web deflection (Section 4.3.2) = $F_y/K_{cw}=41.22 \times 10^3/0.17 \times 10^6$ = 0.246 mm

Endplate deflection (equation 4.1) = $F_y/K_{cf}$ = $41.22 \times 10^3/1.37 \times 10^6$ = 0.0299 mm

Total row 3 deformation: 9.89

Rotation of the connection due to row 3 deformations: $9.89/383 = 0.0289$ mm

Connection rotation = min $(0.0188, 0.023, 0.0289) = 0.0188$

Table 5.1 Summary of calculations of connection component forces at peak beam catenary force

| | Tension | | | | | | | | | | | |
| | Column flange | | Column web | | Endplate | | Beam web | | Bolt tension | | Min. | Effective resistance |
| | Resistance (kN) | Stiffness (N/mm) | Resistance (kN) | Stiffness (N/mm) | Resistance (kN) | Stiffness (N/mm) | Resistance (kN) | Stiffness (N/mm) | Resistance (kN) | Stiffness (N/mm) | (kN) | (kN) |
|---|---|---|---|---|---|---|---|---|---|---|---|---|
| Row 1 alone | 43.73 | 1.01E+06 | 141.63 | 2.42E+05 | 43.73 | 1.45E+06 | N/A | N/A | 43.73 | 1.72E+05 | 43.73 | 43.73 |
| Row 2 alone | 43.73 | 4.58E+05 | 141.63 | 9.85E+04 | 43.73 | 1.37E+06 | 120.96 | 8.36E+04 | 43.73 | 1.72E+05 | 43.73 | 43.73 |
| Row 2 and 1 | 87.46 | | 201.78 | | N/A | | N/A | | | | 87.46 | 43.73 |
| Row 2 | | | | | | | | | | | 43.73 | |
| Row 3 alone | 43.73 | 7.79E+05 | 141.63 | 1.67E+05 | 43.73 | 1.38E+06 | 120.96 | 8.36E+04 | 43.73 | 1.72E+05 | 43.73 | 43.73 |
| Row 3 and 1 and 2 | 128.69 | | 256.44 | | N/A | | N/A | | | | 128.69 | |
| Row 3 | | | | | | | | | | | 41.22 | 41.22 |
| Row 3 and 2 | 87.46 | | 196.40 | | 87.46 | | N/A | | | | 87.46 | |
| Row 3 | | | | | | | | | | | 43.73 | |

| | Compression | | | | | | |
| | Column web | | Beam flange | | | | |
| | Resistance kN | Stiffness N/mm | Resistance kN | Stiffness N/mm | | Min. | Effective resistance |
|---|---|---|---|---|---|---|---|
| Compression | 150.75 | 2.04E+05 | 224.74 | ∞ | | — | 150.75 |

The detailed calculations presented above are for connection component capacities. For extended endplate connections, it is assumed that the two top rows reach their capacities. For other bolt rows, their forces and deformations are calculated based on their rotation when row one is at its maximum deformation.

The total deformation at row 3 is = rotation × lever arm of row 3: 0.0188 × 375 = 7.05 mm

To calculate the corresponding force in this row, the overall load-deflection curve of the row is evaluated as follows.

Initial equivalent stiffness:

$$1/K_{eq} = 1/K_{bolt} + 1/K_{col.flange} + 1/K_{col.web} + 1/K_{endplate} + 1/K_{beam}$$

in which: $K_{bolt} = 8.18 \times 10^6$ N/mm, $K_{col.flange} = 7.79 \times 10^6$ N/mm, $K_{col.web} = 0.17 \times 10^6$ N/mm, $K_{end, plate} = 1.36 \times 10^6$ N/mm, $K_{beam, web} = 0.0836 \times 10^6$ N/mm, $K_{eq}$ their rotation when row one is at = $0.053 \times 10^6$ N/mm

The failure mode is Mode 3 bolt failure. Therefore, the yield and ultimate resistances of the row are as follows:

$$F_y = 0.086 \times 0.9 \times 353 \times 640 = 33.996 \text{ kN},$$

$$F_u = 0.086 \times 0.9 \times 353 \times 800 = 43.72 \text{ kN}$$

Deflection at yield = $33.996 \times 10^3/0.053 \times 10^6 = 0.64$ mm

The maximum deformation of row 3 = 10.75 mm

Load–deformation curve of the assembly of row 3 components:

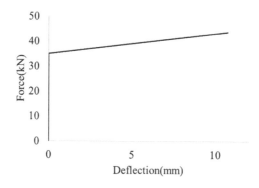

At a deflection of 7.05 mm, the force in row 3 is 40.85 kN.

Beam deflection due to connection rotation (equation 2.1) = $\delta_2 = l \tan (\varphi)/4 = 6000 \tan (0.0188)/4 = 28.2$ mm

Deflection due to thermal elongation (equation 2.6) = $\delta_1 = 2l/\pi\sqrt{(\alpha\Delta\theta + (\alpha\Delta\theta)^2/2)} = (2 \times 6000/\pi)\sqrt{(1.2 \times 10^{-5} \times (742 - 20) + 0.5 \times (1.2 \times 10^{-5} \times (742 - 20))^2)} = 356.39$ mm

Total beam deflection $\delta_{cap}=\delta_1+\delta_2=346.64+28.2=384.54$ mm
$\delta_{peak(demand)}=907.8$ mm
Since $\delta_{cap(supply)}=384.54<\delta_{peak(demand)}$ 907.8 mm, connection failure occurs before the beam reaches the peak catenary action force. Flowchart B should be followed.

### 5.3.4 Flowchart B, finding connection failure temperature before peak catenary force in the beam

Critical temperature for bending: $\theta_{a,\,cr}=584.7°C$
Deflection $\delta_{cr}=(2l/\pi)\ \sqrt{(\alpha\Delta\theta+0.5\ (\alpha\Delta\theta)^2)}$ (equation 2.6)$=(2\times6000/\pi)$ $\sqrt{(1.2\times10^{-5}\times(584.7-20)+0.5\times(1.2\times10^{-5}\times(584.7-20))^2)}=314.96$ mm

Linear interpolation to obtain a trial connection failure temperature
$\theta_{trial}=[(\delta_{cap}-\delta_{low})\ /\ (\delta_{high}-\delta_{low})]\ (\theta_{high}-\theta_{low})+\theta_{low}=[(384.54-314.96)/$ $(907.8-314.96)]\times(742.33-584.7)+584.7\approx603.2°C$
Following the same detailed calculations for connection components as in Section 5.2.3, but at the trial temperature of 603.2°C, the connection rotation is 0.0186. The tension resistance of row 1 is 109.88 kN (Figure 5.9).
The tension force in row 2 is the same as that in row 1: 109.88 kN
Following the same procedure as explained in Section 5.2.3, the load-deflection curve for the assembly of row 3 components at the temperature is constructed from the following values:

$F_y = 87.85$ kN,  $K_{eq} = 0.13\times10^6$,  $\Delta_y = 0.65$ mm,

$F_u = 109.88$ kN,  $\Delta_u = 11.55$ mm.

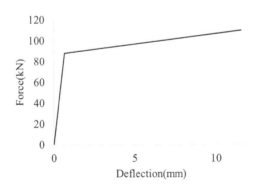

The deflection at row 3=0.0185 × 375=6.93 mm, giving a force of 100.5 kN (see Figure 5.9).
Maximum beam deflection $\delta_{cap}=\delta_1+\delta_2=343.7+9.89=348.21$ mm

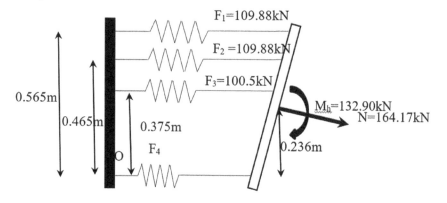

Figure 5.9 shows the state of forces in the connection.

Table 5.2 Summary of results of iterations

| Trial Temp (°C) | Row 1 (kN) | Row 2 (kN) | Row 3 (kN) | N (kN) | $M_h$ (kN·m) | Connection resistance/internal force |
|---|---|---|---|---|---|---|
| 592.5 | 124.41 | 124.41 | 113.77 | 69.58 | 143.09 | 0.97 |
| 590.55 | 127.68 | 127.68 | 116.74 | 52.26 | 144.26 | 0.989 |
| Leaver arm | 0.565 m | 0.465 m | 0.375 m | 0.473 m | – | – |

**Internal forces in the beam at the trial temperature**

$M_s = [-M_p/(\theta_{high} - \theta_{low})] (\theta_{trial} - \theta_{low}) + M_p = [-649 \times 0.518/(742.33 - 584.7)] \times (603.2 - 584.7) + 649 \times 0.518 = 296.72$ kN·m

$M_{h,cr} = M_E - M_s = 486.75 - 336.18 = 150.57$ kN·m

$M_h = [-M_{h,cr}/(\theta_{high} - \theta_{low})] (\theta_{trial} - \theta_{low}) + M_{h,cr} = [-150.57/(742.33 - 584.7)] \times (603.2 - 584.7) + 150.57 = 132.90$ kN·m

$N = [M_E - (M_s + M_h)]/\delta_{trial} = (486.75 - 132.90 - 296.72)/0.348 = 164.17$ kN

The state of forces in the connection are shown in Figure 5.9.

**Check connection resistance:**

Moment resistance of all connection components about the centre of rotation $= 109.88 \times 0.565 + 109.88 \times 0.465 + 0.375 \times 100.5 = 150.8$ kN·m

Moment from beam catenary action force (N) and the beam end moment $(M_h) = 164.17 \times (0.533 - 0.06)/2 + 132.90 = 172.43$ kN·m

150.8 kN·m < 172.43 kN·m → decease the trial temperature.

New trial temperature $= (600.3 + 584.7)/2 = 592.5°C$ (Table 5.2)

## 5.3.5 Flowchart C, finding connection failure temperature after reaching peak catenary force in the beam

In this example, because connection failure is Mode 3 (i.e. bolt failure with no plastic hinges), the connection has very little rotation capacity, making it impossible for the beam to go over the stage of peak catenary force. This is not desirable. However, since the purpose of presenting the worked examples in this chapter is to demonstrate the calculation procedures, in this example of checking connection failure after the beam has passed the peak catenary force, it is assumed that the bolts have much higher resistance and the failure mode of the connection is Mode 1. This is done by assuming that stainless bolts are used with a yield stress of 1700N/mm² and a fracture strain of 0.55.

First trial temperature

$$\theta_{\text{trial}} = \left( \theta_{\text{lower limit}} + \theta_{\text{higher limit}} \right) / 2 = \left( 742.3 + 900 \right) / 2 = 821.15°C$$

Since the top bolt row (row 1) governs connection failure, the calculation results below are for this row only. The same detailed calculations as in Section 5.2.3 are performed, but at a temperature of 821.15°C.

For Mode 1:

Yield resistance: $F_y = 2M_{pl,\,fi}/(m + n) = (2 \times 831.51 \times 10^{-3})/(0.034 + 0.0418) = 22.15$ kN

Ultimate resistance: $F_u = 101.26$ kN (column web in tension)

$\Delta F_{cl} = F_u - F_y = 101.26 - 22.15 = 79.11$ kN        (4.1b)

$E_t = 0.015\ k_E\ E = 0.015 \times 205000 \times 0.085 = 261.38$ N·mm

$L_{\text{eff}} = 40$ mm, $I_c = 2 \times 40 \times 20.5^3/12 = 57434.17$ mm⁴

$\Delta\delta_{cl} = \Delta F/E_t I$ $[((m+k/4)^2\ (m+k/2))/8 - (m+k/4)^3/24 + k^3/1536] = 79.11 \times 10^3/261.38 \times 57434.17[((50+36/4)^2\ (50+36/2))/8 - (50+36/4)^3/24 + 36^3/1536] = 110.97$ mm        (4.1c)

$\Delta_u = \Delta_y + \Delta\delta_{cl} = (22.15 \times 10^3/0.76 \times 10^6) + 110.97 = 110.99$ mm        (4.1a)

$\Delta F_{ep} = F_u - F_y = 101.26 - 19.44 = 81.82$ kN

$L_{\text{eff}} = 40$ mm $I_{ep} = 2 \times 40 \times 25^3/12 = 104166.67$ mm⁴

$\Delta\delta_{ep} = \Delta F/E_t I$ $[((m+k/4)^2\ (m+k/2))/8 - (m+k/4)^3/24 + k^3/1536] = 96 \times 10^3/261.38 \times 104166.67$ $[((50+36/4)^2\ (50+36/2))/8 - (50+36/4)^3/24 + 36^3/1536] = 63.29$ mm

$\Delta_u = \Delta_y + \Delta\delta_{cl} = (19.44 \times 10^3/1.09 \times 10^6) + 63.29 = 63.31$ mm        (4.1a)

Bolt deflection = $0.55 \times 52.5 = 26$ mm (it is assumed that the bolts reach their maximum deflection capacity.)

Deformation on the compression side:

$\Delta_u = (l_y t_{wc}^2)/(2b_{fc} t_{fc}) \sqrt{((t_{wc} d_{wc})/(t_{fc}c))} \chi_F$ (4.16)

$c = t_{fb} + \sqrt{2} a_{ep} + t_{ep} + \min(u, \sqrt{2} a_{ep} + t_{ep}) = 25 + 10 + 10\sqrt{2} + 15.6 = 64.74$ mm
(4.15)

$m_1 = f_{yc}b_{fc}/f_{y, wc}t_{wc} = 345 \times 258.8/(355 \times 12.8) = 19.65,$

$m_2 = 0.02(b_{wc}/t_{fc}) = 0.02 \times (200.3/20.5) = 0.195$

$l_y = c + 2t_{fc} (1 + \sqrt{(m_1 + m_2)}) = 64.74 + 2 \times 20.5(1 + (0.195 + 19.65)^{0.5}) = 288.39$ mm

$\lambda_F = \sqrt{((l_y t_{wc} k_y f_{y, wc})/(0.9 k_F k_E E t_{wc}^3/d_{wc})}$
   $= ((288.39 \times 12.8 \times 275 \times 0.099)/(0.9 \times 6 \times 205000 \times 12.8^3 \times 0.085))^{0.5}$
   $= 0.319$

$\chi_F = 0.06 + (0.47/0.319) = 1.53$

$\Delta_u = (l_y t_{wc}^2)/(2b_{fc}t_{fc})\sqrt{((t_{wc}d_{wc})/(t_{fc}c))}\chi_F = (288.39 \times 12.8^2)/(2 \times 258.8 \times 20.5)$
   $[(12.8 \times 200.3/20.5 \times 64.74)^{0.5}] = 9.46$ mm

Total deflection $= 26 + 63.29 + 110.99 + 9.46 = 209.74$ mm

$\varphi = 209.74/565 = 0.37$ rad

$\delta_2 = l \tan (\varphi)/4 = 6000 \tan (0.373)/4 = 586.98$ mm

$\delta_1 = 2l/\pi\sqrt{(\alpha\Delta\theta + (\alpha\Delta\theta)^2/2)} = (2 \times 6000/\pi)\sqrt{(1.2 \times 10^{-5} \times (821.5 - 20) + 0.5 \times (1.2 \times 10^{-5} \times (821.5 - 20))^2)} = 375.50$ mm

$\delta_{trial} = 586.98 + 375.50 = 962.48$ mm

Following the same detailed calculations as in Section 5.2.3, the force–deformation curve of the assembly of row 3 is obtained as shown in Figure 5.10. And the internal forces in different connection components are:

**Catenary force at the trial temperature**

$N = M_E/\delta_{trial} = 486.75/0.962 = 505.98$ kN

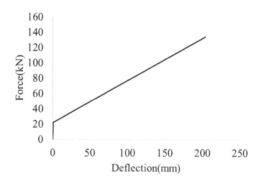

*Figure 5.10* The force–deformation curve of the assembly of row 3; Row 1 = 101.26 kN, Row 2 = 101.26 kN, Row 3 = 86.04 kN.

### Check connection resistance

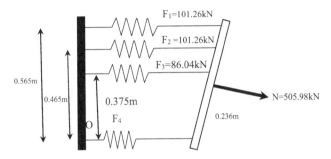

Figure 5.11 Shows all the internal forces on the connection.

Table 5.3 Summary of calculation results of iterations

| Trial Temp (°C) | Row 1 (kN) | Row 2 (kN) | Row 3 (kN) | N (kN) | Connection resistance/ internal force |
|---|---|---|---|---|---|
| 860.58 | 81.32 | 81.32 | 60.71 | 482 | 0.83 |
| 840.86 | 91.38 | 91.38 | 67.56 | 493.37 | 0.93 |
| 831.2 | 96.32 | 96.32 | 70.91 | 499.1 | 0.98 |
| 826.35 | 98.79 | 98.79 | 72.59 | 502.04 | 1.004 |
| Lever arm | 0.565 m | 0.465 m | 0.375 m | 0.473 m | – |

### Check connection resistance

Moment resistance of all connection components about the centre of rotation (Figure 5.11).

$$=101.26 \times 0.565 + 101.26 \times 0.465 + 0.375 \times 86.04 = 151.9 \text{ kN·m}$$

Moment from beam catenary action force $(N)$ = 505.98 × 0.533/2 = 134.84 kN·m

151.9 kN·m > 134.84 kN·m → increase the trial temperature.

For the next trial, the $\theta_{\text{lower limit}}$ will be 821.15°C, giving a new trial temperature of (821.15+900)/2=860.58°C. Table 5.3 summarises the key calculation results of the subsequent iterations.

## REFERENCE

Brown, D., Iles, D., Brettle, M. & Malik, A. 2013. Joints in Steel Construction: Moment-Resisting Joints to Eurocode 3. BCSA/SCI Connections Group.

# Chapter 6

# Hollow section connections

## 6.1 INTRODUCTION

Steel tubular sections include square/rectangular hollow sections (SHS/RHS), circular hollow sections (CHS) and elliptical hollow sections (EHS). Owing to their many attractions, including high structural efficiency (high-strength-to-weight ratio) and attractive appearance, they are commonly used in construction.

Application of steel tubular sections in construction can be either in whole tubular structures such as trusses or as columns (either hollow or concrete filled) in beam-to-column construction. Therefore, connections to tubular members vary depending on the type of construction. In general, they can be classified into two categories: welded tubular connection (Figure 6.1a) with the tubular members predominately in axial tension or compression, or steel beam to tubular column connection (Figure 6.1b) where bending, shear and axial force may transfer from the beam to the steel or concrete filled tubular column. The general method described in this book so far can be directly applied to steel beam to tubular column connections. However, research is still ongoing to develop component behaviour models for steel beam to column tubular connections. Because of this, it is not covered in this book.

This book will only cover welded tubular connections with the connected members predominantly in axial tension or compression, as in truss construction. Design for this type of application at ambient temperature is currently well covered in EN1993-1-8 (CEN, 2005) and CIDECT Design guides (Wardenier et al., 2002, 2008, Packer et al., 2009, Rondal, 1992).

However, there are two issues with direct application of the same ambient temperature design approach to fire design even with modification of steel properties:

1. when calculating welded tubular connection load-carrying capacity, the ambient temperature equations are a function of the yield strength of steel. Since the effects of stiffness are involved, and Young's modulus of steel decreases faster than the yield strength of steel, it may not

DOI: 10.1201/9781003134466-6

(a)

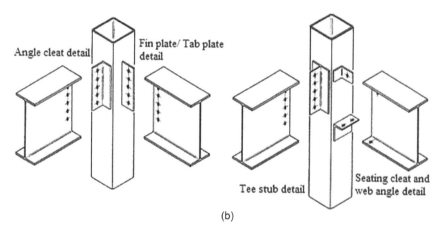

(b)

*Figure 6.1* Connections for constructions using steel tubular elements. (a) Welded tubular connection ( HYPERLINK \l "Wang and Ozyurt, 2021). (b) Steel beam to tubular columns ( HYPERLINK \l "France et al., 1999).

be safe to replace the yield stress of steel at ambient temperature with that at elevated temperature.

2. The global deflection of welded tubular trusses changes the configuration of truss members, thereby inducing additional forces in some brace members.

These topics have been systematically investigated by Ozyurt (2015), Ozyurt and Wang (2015, 2016) and Wang and Ozyurt (2021) for different types of tubular connections. In fact, a substantial amount of their findings on welded steel tubular connection load-carrying capacity has been adopted by the forthcoming new version of Eurocode EN 1993-1-2. This short chapter will summarise their main findings.

## 6.2 LOAD-CARRYING CAPACITY OF WELDED TUBULAR CONNECTIONS

Figure 6.2 shows the failure modes of welded tubular connections according to EN 1993-1-8 (CEN, 2005). They are as follows:

a. Chord face failure (plastic failure of the chord face) or chord plastification (plastic failure of the chord cross-section)
b. Chord side wall failure (or chord web failure) by yielding, crushing or instability (crippling or buckling of the chord side wall (or chord web) under compression in the brace member
c. Chord shear failure
d. Punching shear failure of a hollow section chord wall (crack initiation leading to rupture of the brace member from the chord member)
e. Brace failure with reduced effective width (cracking in the welds or in the brace members)
f. Local buckling failure of the brace member or of the hollow section chord member at the connection location

For chord shear failure (failure mode c), punching shear failure of the hollow section chord wall (failure mode d), brace failure by cracking in the weld or in the brace member (failure mode e) and local buckling failure of the brace member (failure mode f), they are all controlled by the yield strength of steel. Therefore, the resistance of welded tubular connection for these failure modes at elevated temperatures can be calculated by simply multiplying the ambient temperature resistance of the connection by the steel yield strength reduction factor at the elevated temperature.

For chord side wall failure (failure mode b), the ambient temperature resistance equation contains both the yield strength and Young's modulus of steel to allow for yield, crushing or instability (buckling or crippling) of the side wall. Therefore, for extension to elevated temperatures, it is only necessary to replace the ambient temperature values of Young's modulus and yield strength of steel by the corresponding values at elevated temperatures.

The joint resistance calculation equations for chord face failure (failure mode a) are based on the formation of a yield line mechanism on the surface of the chord face. At elevated temperatures, the chord face may undergo large deformations thereby modifying the geometric shape of the yield line mechanism. Furthermore, the ambient temperature resistance calculation equations are modified, when necessary, to limit the maximum deflection of the chord face to 3% of the chord width. Both these aspects indicate that the stiffness of the chord face has some influence on the chord face resistance. However, the ambient temperature resistance calculation equations are only a function of the yield strength of steel. Therefore, it can be considered that the ambient temperature equations have embedded a fixed,

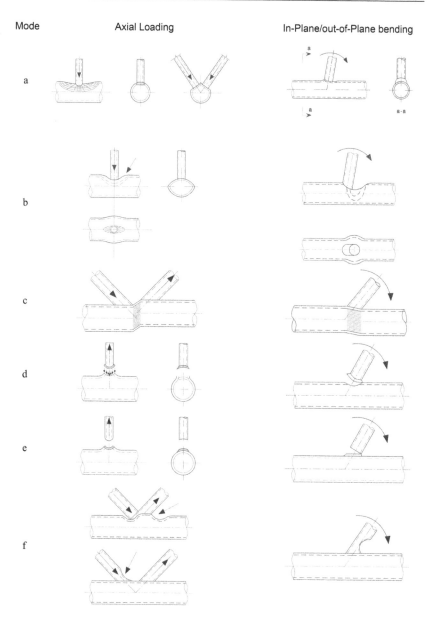

*Figure 6.2* Possible failure modes of welded tubular connections (CEN, 2005).

but unknown, ratio of steel yield strength to Young's modulus. At elevated temperatures, the same implied ratio in these equations, based on the ambient temperature of Young's modulus and yield strength of steel, can no longer be applied. In particular, since Young's modulus of steel decreases more than the yield strength of steel at elevated temperatures, it may not be safe ignoring the effects of Young's modulus decreasing faster than the yield strength of steel by multiplying the ambient temperature equations by the yield strength reduction factor only.

Since it is not feasible to attract sufficient resources to tackle this problem to produce exact solutions, the research studies of Ozyurt et al. (2014), Ozyurt and Wang (2015, 2016), Wang and Ozyurt (2021) were conducted to develop a simple and safe method to address the above concern for different connection arrangements under different loading conditions.

## 6.2.1 Brace member in axial load

Based on extensive numerical simulations for different types of planar welded tubular truss connections under different brace-loading conditions, they have reached the following conclusions:

1. When the brace members in T-, Y- and X- connections with SHS/ RHS, CHS sections EHS sections in type-1 and type-2 arrangement (connection to the wide side of EHS section) as shown in Figure 6.4 are in compression, the formulations in EN1993-1-8 (CEN, 2005) can be utilised at elevated temperature, only if the ambient temperature resistance is modified by multiplying the elevated temperature reduction factor for Young's modulus of steel, rather than the reduction factor for the yield stress of steel.
2. In other cases, the formulations given in EN1993-1-8 (CEN, 2005) for ambient temperature design can be employed for elevated temperature calculations provided that elevated temperature steel yield stress is used.

Multiplanar DKK connections, as shown in Figure 6.3, represent a special case. Interested readers should consult the work of Wang and Ozyurt (2021).

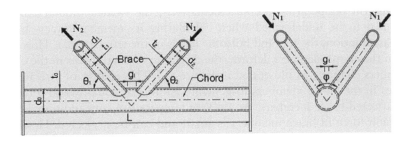

*Figure 6.3* DKK configuration.

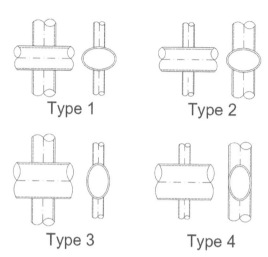

*Figure 6.4* Elliptical hollow section elements in X-type connection (Ozyurt and Wang, 2018).

## 6.2.2 Brace members in bending

Ozyurt and Wang (2018) extended their research to planar welded T-, Y-, K- and X- truss connections using CHS, RHS and EHS when the bracing members are under bending. Their main conclusion is when the brace member is under in- or out-of- plane bending moment, half of the chord face undergoes longitudinal tension and the other half undergoes compression. Because of this, the elevated temperature load-carrying capacity of the connection can be calculated by modifying that of the connection at ambient temperature by multiplying the average of the modification factors in the above section for the brace member under pure tension and pure compression.

## 6.3 INCREASE IN BRACE MEMBER FORCE IN FIRE

In ambient temperature design calculations, the welded tubular truss is assumed to be undeformed when calculating its member forces. Under this assumption, the top and bottom chords of the truss are flat. However, when the chord members deform, the direction of axial forces in the chord members changes, as illustrated in Figure 6.5. This change of direction of forces in the chord members to generate additional compression forces in the affected brace members. At ambient temperature, any additional compression force in brace members is small and can be neglected. However, at high temperatures in fire, the truss deflection can be large and it may no longer be safe to ignore the additional forces in brace members.

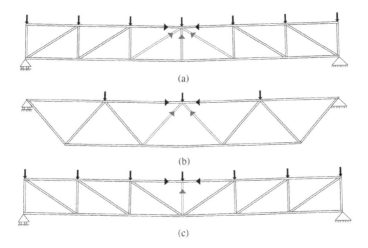

*Figure 6.5* Effects of truss deformation on additional compression force in brace members (from Ozyurt and Wang 2015). (a) Howe Truss. (b) Warren Truss. (c) Pratt Truss.

The additional compressive force in brace member is the highest where the angle between forces in the chord on either side of the brace member is the largest, which coincides with the location of the maximum deflection of the truss. In brace members further away from this brace member, the additional compressive force rapidly decreases.

Ozyurt and Wang (2015) suggest the following simplified equations to calculate the additional compressive force in brace members:

The total compressive force in the brace member at the centre of Warren truss:

$$F_{\text{truss-centre}} = F_{\text{truss-centre, 0}} + \frac{F_{\text{maximum chord compression}}\delta}{L\,\text{Sin}\theta} \tag{6.1}$$

The total compressive force in Warren truss brace members other than the centre brace member:

$$F_{\text{other brace member}} = F_{\text{other brace member, 0}} + \frac{F_{\text{maximum chord compression}}\delta}{L\,\text{Sin}\theta}\left(\frac{d}{L/2}\right) \tag{6.2}$$

where

d is the distance along the chord from the support to the brace member of interest,

$F_{\text{truss-centre,0}}$ is the compression force in the centre brace member at ambient temperature under the fire limit state design loads,

$F_{\text{maximum chord compression}}$ is the maximum compressive force in the chord member at the truss centre at ambient temperature under the fire limit state design loads,

$F_{\text{truss-centre}}$ is the total compressive force in the brace member at the truss centre for fire design,

$F_{\text{other brace member}}$ is the total compressive force in other brace members for fire design,

$L$ is the overall span of the truss,

$\delta$ is the maximum deflection of the truss in fire, which may be taken as $L/30$,

$\theta$ is the angle between the brace member and the compressive chord.

For Pratt truss, since there is only one brace member at the middle of the truss and $\theta = 90°$, equation 6.1 should be modified to

$$F_{\text{truss-centre}} = F_{\text{truss-centre, 0}} + \frac{2F_{\text{maximum chord compression}}\,\delta}{L} \tag{6.3}$$

### 6.3.1 Worked example

A worked example is presented below to briefly demonstrate the calculation method for additional compressive forces due to truss deflection in fire. Details of the truss are in Figure 6.6.

Due to symmetry, Figure 6.7 shows results of internal member forces for the left half of the truss, where the negative sign indicates compression.

**Calculations of additional and total compressive forces in compressive brace members**

For the centre brace member (No. 11 in Figure 6.6):

$\Theta = 38.7°$, $\delta = 36000/30 = 1200$ mm.

Equation 6.2 gives

$$F_{11} = -51.6 + \left(-688.8 \times 1200 / \left(36000 \times \text{Sin}(38.7)\right)\right) = -88.32\,\text{kN}$$

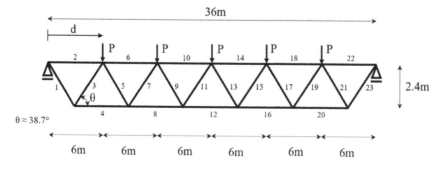

*Figure 6.6* Details of an example Warren-type truss with K joints (from Wardenier et al., 2008).

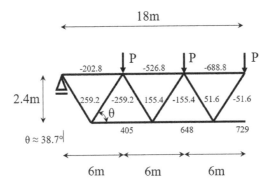

Figure 6.7 Member internal forces without considering truss deflection.

For other compressive brace members:

Member No. 7:
$d$ =12000 mm,

Equation 6.3 gives $F_7 = -155.4 + (-526.8 \times 1200/(36000 \times Sin\ (38.7))(12000/18000) = -174.12$ kN

Member No. 3:
$d$=6000 mm,

Equation 6.3 gives: $F_3 = -259.2 + (-202.8 \times 1200/(36000 \times Sin\ (38.7))(6000/18000) = -262.80$kN

Table 6.1 compares the internal forces in these three compressive brace members between without and with consideration of truss deflection. Because the compressive force in a brace member increases as it moves away from the centre while the additional compressive force in the same member decreases rapidly, the percentage increase in compressive force in brace members is very low except for the centre and the immediate adjacent compressive brace members. Therefore, in practical fire design, only these members need checking.

The results in Table 6.1 also indicate that the centre brace member has a very low compressive force, as expected due to small shear forces near the centre of the truss. Therefore, if the centre brace member is constructed in the same way as other brace members, even the additional compressive force in the member would not cause any problem to fire resistance of the truss. However, if the centre brace member and its connection are to be sized according to the force in the brace member, then the additional compressive force in the member should be included.

*Table 6.1* Comparison of internal forces in compressive brace members

| Member No. | Without considering truss deflection from ambient temperature calculation | With consideration of truss deflection for fire design (% change) |
|---|---|---|
| 3 | −259.2 | −262.8 (1.4%) |
| 7 | −155.4 | −174.12 (12%) |
| 11 | −51.6 | −88.32 (71.2%) |

## REFERENCES

CEN 2005. EN 1993-1-8 Eurocode 3: Design of steel structures, Part 1–8: Design of joints. *British Standard.*

France, J. E., Buick Davison, J. & Kirby, P. A. 1999. Strength and rotational stiffness of simple connections to tubular columns using flowdrill connectors. *Journal of Constructional Steel Research,* 50, 15–34.

Ozyurt, E., 2015. *Behaviour of Welded Tubular Structures in Fire.* The University of Manchester, United Kingdom.

Ozyurt, E. & Wang, Y. 2016. Effects of non-uniform temperature distribution on critical member temperature of steel tubular truss. *Engineering Structures,* 116, 95–106.

Ozyurt, E. & Wang, Y. C. 2015. Effects of truss behaviour on critical temperatures of welded steel tubular truss members exposed to uniform fire. *Engineering Structures,* 88, 225–240.

Ozyurt, E. & Wang, Y. C. 2018. Resistance of axially loaded T-and X-joints of elliptical hollow sections at elevated temperatures–a finite element study. *Structures,* 14, 15–31.

Ozyurt, E., Wang, Y. C. & Tan, K. H. 2014. Elevated temperature resistance of welded tubular joints under axial load in the brace member. *Engineering Structures,* 59, 574–586.

Packer, J. A., Wardenier, J., Zhao, X.-L., Van der Vegte, G. & Kurobane, Y. 2009. *Design guide for rectangular hollow section (RHS) joints under predominantly static loading,* Cidect.

Rondal, J. 1992. *Structural Stability of Hollow Sections,* Koln, Verlag TÜV, Rheinland.

Wang, Y. & Ozyurt, E. 2021. Static resistance of axially loaded multiplanar gap KK-joints of Circular Hollow sections at elevated temperatures. *Engineering Structures,* 229, 111676.

Wardenier, J., Kurobane, Y., Packer, J. A., Van der Vegte, G. & Zhao, X. 2008. *Design guide for circular hollow section (CHS) joints under predominantly static loading,* Cidect.

Wardenier, J., Packer, J., Zhao, X.-L. & Van der Vegte, G. 2002. *Hollow Sections in Structural Applications,* Bouwen met staal Rotterdam, The Netherlands.

# Chapter 7

# Methods of improving connection performance

## 7.1 INTRODUCTION

Ordinarily, explicit calculations as detailed in this book are not required when checking fire resistance of connections, on the conditions that the connection is protected to the maximum fire protection thickness of the connected members and the maximum load ratio in any connection component does not exceed that of the corresponding member. This is based on the fact that the connection will have lower temperatures than the connected members due to its higher thermal mass (or lower section factor). This is the fire-resistant (FR) design recommendation in EN 1993-1-2 (CEN, 2005). However, this practice is intended for the fire limit state design of bending in the connected beams (as indicated by stage 2 in Chapter 2). Before this bending limit, connection failure is unlikely as has been explained in Chapter 2. Therefore, explicit checking of connections in fire is most likely needed only when the structure has to survive after the bending limit of the beams. This situation arises when dealing with the control of disproportionate/progressive collapse of the structure in fire.

As with the control of disproportionate collapse in the cold condition, ensuring structural integrity (robustness) in fire after having reached its fire limit state of bending involves exploitation of any additional load-carrying capacities of the structure or alternative load paths of the structure, in the following two generic ways:

1. It is common practice that when checking for fire resistance of a structure, connections are assumed to be pinned. In reality, all connections have some bending moment resistance. Therefore, any real bending moment resistance of the connection can be used in checking robustness of the structure.
2. The connected beams may go into catenary action which can generate much greater loadbearing capacity than under bending.

However, using method (1) above is unlikely to be sufficient. Consider the situation of column removal, which is often the scenario in the context of

DOI: 10.1201/9781003134466-7

*Figure 7.1* A steel beam in catenary action in fire (Wang et al., 2010).

checking structural robustness. Assuming equal beam span on both sides and the connections being able to provide full strength and total continuity, the maximum bending moment in the connection of the double span beam would be four times that in the single span beam as the maximum bending moment is proportional to span squared. Even if the connections are able to provide full strength of the connected beam and this full strength is not utilised in normal fire limit state design for bending, the design resistance of the structure is at most doubled, which would still not be adequate to resist four times the internal force.

Using method (2) is more achievable as the load-carrying capacity of the structure can be controlled by deflection of the structure. However, in order for catenary action to fully work, the connections of the structure should be able to develop very large rotations without fracture while maintaining the tensile resistance of the connected beams, as shown in Figure 7.1.

The rotation capacity for connections required to allow substantial development of catenary action in the connected beams is in the order of 20° (Wang et al., 2010). Unfortunately, commonly used steel and composite beam-column connections are not able to achieve this rotation capacity, with a rotation capacity of less than 10° and typically about 5°, as shown in Table 7.1 based on the experimental results of Yu et al. (2009).

Therefore, research studies are being conducted to develop connections that are able to develop the required large rotation capacity. These research studies can be grouped into two categories: (1) using ductile steels for key connection components and (2) developing new connection components. The following two sections present a brief introduction to some of these developments.

*Table 7.1* Summary of the results reported by Yu et al. (2009)

| Connection | Specimen geometry | Temperature (°C) | Applied load angle α (°) | Force (kN) | Rotation (°) | Failure mode |
|---|---|---|---|---|---|---|
| Fin plate | 3-8.8-20 | 20 | 53.85 | 145.95 | 8.107 | Bolt shear |
| Fin plate | 3-8.8-20 | 450 | 51.47 | 70.48 | 6.093 | Bolt shear |
| Fin plate | 3-8.8-20 | 550 | 53.44 | 34.81 | 6.558 | Bolt shear |
| Fin plate | 3-8.8-20 | 650 | 53.09 | 17.99 | 6.255 | Bolt shear |
| Partial endplate | 3 bolt rows | 20 | 55 | 179.3 | 11.2 | Plate fracture |
| Partial endplate | 3 bolt rows | 450 | 55 | 55.6 | 4.8 | Plate fracture |
| Partial endplate | 3 bolt rows | 550 | 55 | 36.3 | 3.9 | Plate fracture |
| Partial endplate | 3 bolt rows | 650 | 55 | 22.1 | 4.5 | Plate fracture |
| Web cleat | 3-8.8-20 | 20 | 55 | 186.34 | 16.57 | Punching shear |
| Web cleat | 3-8.8-20 | 450 | 55.8 | 93.74 | 9.39 | angle fracture |
| Web cleat | 3-8.8-20 | 550 | 56 | 52.91 | 10.52 | angle fracture |
| Web cleat | 3-8.8-20 | 650 | 56.5 | 25.7 | 14.15 | bolt shear |
| Flush endplate | Plate thickness 10 mm | 20 | 43.8 | 259 | 4.4 | Plate fracture |
| Flush endplate | Plate thickness 10 mm | 450 | 46.7 | 182.5 | 5.6 | Plate fracture |
| Flush endplate | Plate thickness 10 mm | 550 | 47.2 | 87.8 | 2.2 | Bolt fracture |
| Flush endplate | Plate thickness 10 mm | 650 | 48.1 | 39.2 | 3.1 | Bolt fracture |
| Flush endplate | Plate thickness 10 mm | 550 | 35 | 105.9 | 2.2 | Bolt fracture |
| Flush endplate | Plate thickness 8 mm | 550 | 35 | 101 | 7.5 | Bolt fracture |
| Flush endplate | Plate thickness 15 mm | 550 | 35 | 124.6 | 2.5 | Bolt fracture |

## 7.2 USING DUCTILE CONNECTION COMPONENTS

The modest rotation capacity of common beam to column connections is a result of the limited maximum elongation of carbon steel at ambient and elevated temperatures, being less than 20% (CEN, 2005) and even lower for carbon steel bolts. Two other types of steel can achieve much higher elongation at elevated temperatures: stainless steel and FR steel. For example, Sakumoto et al. (1993) report that FR steel bolts can reach elongations of 34% and 57% at 650°C and above 700°C, respectively. According to Gardner (2005), the maximum elongations (strain) of stainless steel are 40%–60%.

Elsawaf and Wang (2012) and Chen and Wang (2012) have numerically investigated the effects of using FR bolts instead of high-strength carbon steel bolts of a steel beam connected by endplate connections. Figure 7.2 shows one set of the results of Chen and Wang (2012) using M20 G10.9 carbon steel bolts (Group 3 results in Figure 7.2), the steel is able to develop a very limited amount of catenary action, allowing the beam to survive less than 50°C beyond the bending limit (which is indicated by zero axial force in Figure 7.2). However, if using the same size and grade of FR bolts (Group 6 and Group 9, green line results in Figure 7.2), the connected beam is able to survive almost a further 200°C after reaching the bending limit state.

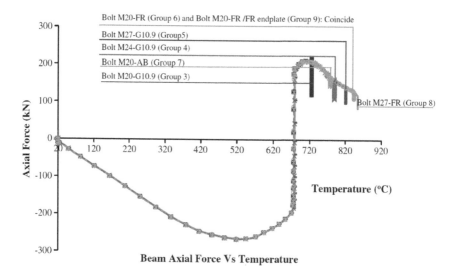

*Figure 7.2* Comparison of beam axial load-temperature variation (tension positive) using different types and grades of bolts (Chen and Wang, 2012).

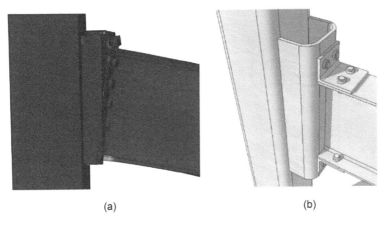

(a)                                                    (b)

*Figure 7.3* Reverse channel connections with different detailing. (a) Reverse channel with extended endplate (Elsawaf and Wang, 2012). (b) Reverse channel with top and seat angle connections (Málaga-Chuquitaype and Elghazouli, 2010).

## 7.3 DEVELOPING MORE DUCTILE CONNECTION COMPONENTS

While it is possible to use the more ductile FR steel or stainless steel in key connection components (e.g. bolts/endplate) to improve connection behaviour (mainly rotation capacity) in fire, these solutions could be costly. A more feasible approach is to develop new connection components that are able to fulfil their cold design requirements but have much better performance in fire. A few of these developments are highlighted in this section.

### 7.3.1 Reverse channel connection

Figure 7.3 shows an example of connection using reverse channel. This connection type was originally developed to overcome the difficulty of making semi-continuous connections to tubular columns due to the problem of access from inside the tube for bolted endplate connection or lack of strength and rigidity of fin plate connection. By using a channel between the beam and the column, bolting endplate can be easily accessed from the space enclosed by the short channel section, and welding two channel legs near the steel tube walls provides much improved strength and stiffness than using one fin plate in the middle of the steel tube where it is most flexible and weak.

A number of research studies have confirmed the advantages of this type of connection in terms of its improved bending moment –rotation behaviour, as exemplified in Figure 7.4 (Ding and Wang, 2007, Málaga-Chuquitaype and Elghazouli, 2010, Huang et al., 2013, Lopes et al., 2015, Wang and Xue, 2013, Jafarian and Wang, 2015a, b, 2016).

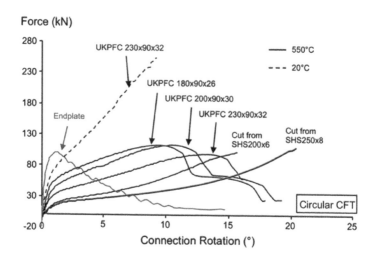

Figure 7.4 Comparison between load-rotation behaviour of endplate and reverse channel connection at elevated temperature (Huang et al., 2013).

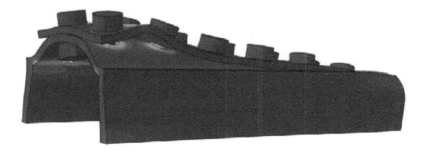

Figure 7.5 Potential deformation capability of reverse channel (Elsawaf and Wang, 2012).

The channel section has a further attractive feature that can be explored in improving steel structural resistance to disproportionate collapse in fire: the channel section could be designed to provide a high level of rotation capacity to the connection, as shown in Figure 7.5.

Although the reverse channel connection was developed for tubular columns, it is equally applicable to improving the rotation capacity of connections to open section columns.

## 7.3.2 Semi-circular web cleat

The proposed connection method by Liu et al. (2020), shown in Figure 7.6, may be considered an important development tracing its root to reverse channel connection outlined in the previous section. As shown in Figure 7.6a, by

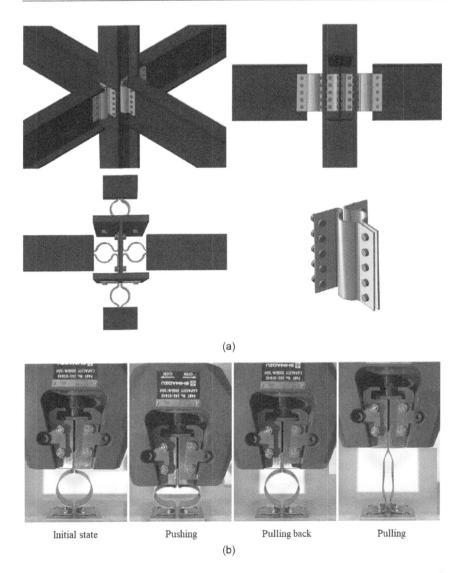

(a)

(b)

Initial state     Pushing     Pulling back     Pulling

*Figure 7.6* Modified web cleat connection (Liu et al., 2020, 2021). (a) Schematic detail of the connection. (b). Test on thin cold rolled mokup of the connection.

using semi-circular web cleats, the channel shape is changed to a spherical shape, thus providing great rotation capacity for the connection by straightening of the circular shape, as shown in Figure 7.6b.

The results of Liu et al. (2020) suggest that this type of connection is able to achieve 0.35 radian rotation at the point of failure, shown as "novel connection" in Figure 7.7.

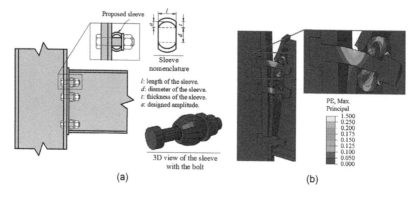

*Figure 7.7* Comparison between the rotational capacity of the semi-circular web cleat connection and other types of connections (Liu et al. 2020).

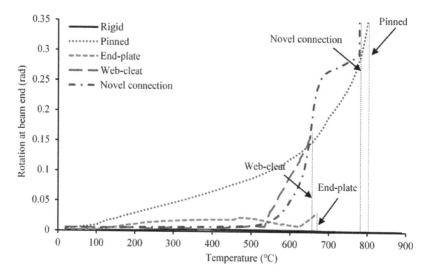

*Figure 7.8* Bolt sleeve solution for endplate connection (Shaheen et al., 2022).

### 7.3.3 Bolt sleeve

Bolt sleeve has been recommended by Shaheen et al. (2022) as a solution to improve the ductility of endplate connection, see Figure 7.8. As shown in Figure 7.8, in this concept the detailing of the endplate connection is not changed. Instead, a steel sleeve, which will have lower capacity than bolts, is placed between the bolt head and the endplate. As illustrated in Figure 7.8, the sleeve is activated when the connection goes under tension and this achieves increased ductility of the endplate by the sleeve getting compressed between the washer and endplate.

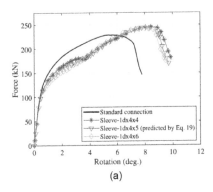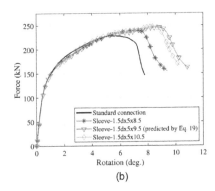

(a)                                    (b)

*Figure 7.9* Comparison between the load-rotation curves of endplate connection with and without bolt sleeve (Shaheen et al., 2022). (a) Sleeve with $L/d_b = 1.0$ and $t = 4\,mm$. (b) Sleeve with $L/d_b = 1.5$ and $t = 5\,mm$.

Some results of Shaheen et al. (2022), extracted in Figure 7.9, suggest that this solution could potentially improve the ductility of an endplate up to 50%.

## 7.4 SUMMARY

Connections have always been the key components of structures. Their role in preventing disproportionate/progressive collapse of structures in fire is even more prominent, and much improvement is needed to allow connections to fulfil the role. Among the two options of increasing the strength (bending moment resistance) and rotation capacity of connections, the latter is more effective. This can be achieved either by using FR steel or stainless steel to replace carbon steel in the most critical connection components (bolts or endplate), or by using connection components that have inherently more ductile behaviour with reverse channel, semi-circular web cleat and bolt sleeve being examples. However, much further research is needed to demonstrate their effectiveness in complete and full-scale steel structures in fire.

## REFERENCES

CEN 2005. EN 1993-1-2 Eurocode 3: Design of Steel Structures–Part 1–2: General Rules–Structural Fire Design. *British Standards Institution*.
Chen, L. & Wang, Y. 2012. Methods of improving survivability of steel beam/column connections in fire. *Journal of Constructional Steel Research*, 79, 127–139.

Ding, J. & Wang, Y. C. 2007. Experimental study of structural fire behaviour of steel beam to concrete filled tubular column assemblies with different types of joints. *Engineering Structures*, 29, 3485–3502.

Elsawaf, S. & Wang, Y. C. 2012. Methods of improving the survival temperature in fire of steel beam connected to CFT column using reverse channel connection. *Engineering Structures*, 34, 132–146.

Gardner, L. 2005. The use of stainless steel in structures. *Progress in Structural Engineering and Materials -Wiley Online Library*, 7, 45–55.

Huang, S.-S., Davison, B. & Burgess, I. W. 2013. Experiments on reverse-channel connections at elevated temperatures. *Engineering Structures*, 49, 973–982.

Jafarian, M. & Wang, Y. 2015a. Tying resistance of reverse channel connection to concrete filled square and rectangular tubular sections. *Engineering Structures*, 100, 17–30.

Jafarian, M. & Wang, Y. C. 2015b. Force–deflection relationship of reverse channel connection web component subjected to transverse load. *Journal of Constructional Steel Research*, 104, 206–226.

Jafarian, M. & Wang, Y. C. 2016. Elastic stiffness of reverse channel joint web components under bolt tension. *Archives of Civil and Mechanical Engineering*, 16, 961–981.

Liu, Y., Huang, S.-S. & Burgess, I. 2020. Performance of a novel ductile connection in steel-framed structures under fire conditions. *Journal of Constructional Steel Research*, 169, 106034.

Liu, Y., Huang, S.-S. & Burgess, I. 2021. Ductile connection to improve the fire performance of bare-steel and composite frames. *Journal of Structural Fire Engineering*, 13, 249–266.

Lopes, F., Santiago, A., Da Silva, L. S., Iqbal, N., Veljkovic, M. & Da Silva, J. G. S. 2015. Sub-frames with reverse channel connections to CFT composite columns-experimental evaluations. *Advanced Steel Construction*, 11, 110–125.

Málaga-Chuquitaype, C. & Elghazouli, A. 2010. Behaviour of combined channel/angle connections to tubular columns under monotonic and cyclic loading. *Engineering Structures*, 32, 1600–1616.

Sakumoto, Y., Keira, K., Furumura, F. & Ave, T. 1993. Tests of fire-resistant bolts and joints. *Journal of Structural Engineering*, 119, 3131–3150.

Shaheen, M. A., Foster, A. S. & Cunningham, L. S. 2022. A novel device to improve robustness of end plate beam–column connections: Analytical model development. *Thin-Walled Structures*, 172, 108878.

Wang, Y., Davison, J., Burgess, I., Plank, R., Yu, H., Dai, X. & Bailey, C. 2010. The safety of common steel beam/column connections in fire. *Structural Engineer*, 88, 26–35.

Wang, Y. & Xue, L. 2013. Experimental study of moment–rotation characteristics of reverse channel connections to tubular columns. *Journal of Constructional Steel Research*, 85, 92–104.

Yu, H., Burgess, I. W., Davison, J. B. & Plank, R. J. 2009. Experimental investigation of the behaviour of fin plate connections in fire. *Journal of Constructional Steel Research*, 65, 723–736.

# Appendix A: Thermal and mechanical properties of different types of steel at elevated temperatures

## A.1 INTRODUCTION

This appendix provides data for the thermal and mechanical properties of different types of steel at elevated temperatures, including carbon steel, stainless steel, bolts and welds. These data are extracted from EN1993-1-2 (CEN, 2005). This appendix also includes mechanical properties of fire-resistant (FR) steel and FR bolts, extracted from Kelly and Sha (1999) for FR steel and Sakumoto et al. (1993) for FR steel bolts.

The thermal and mechanical properties of steel are temperature dependent. In the expressions of this chapter, steel temperature is denoted by $\theta_a$. Unless otherwise the unit of temperature is °C.

### A.1.1 Thermal properties

The thermal properties include emissivity, thermal conductivity, specific heat capacity and thermal expansion.

### A.1.2 Emissivity

The emissivity of a material expresses the ability of its surface in absorbing radiant heat. It is the ratio of the radiative heat absorbed by the surface to that of an ideal black body surface (CEN, 2002). Based on this definition, the emissivity of an ideal black body is 1.0. As summarised in Table A.1, the emissivity of carbon steel can be taken as 0.7. Stainless steel reflects a large percentage of radiative heat so that a low emissivity value of 0.4 can be used. Hot Dip Galvanised steel has a layer of zinc on the surface, which is highly reflective, but melts away at temperature of about 500°C, after which the emissivity of its surface returns to that of carbon steel.

### A.1.3 Thermal conductivity

Thermal conductivity expresses the ability of a material in conducting heat through the material. This value depends on the material type and

Table A.1  Emissivity of different types of steel

| Type of steel | Emissivity (≤500°C) | Emissivity (>500°C) |
|---|---|---|
| Carbon steel | 0.7 | |
| Stainless steel | 0.4 | |
| Hot-dip galvanized steel[a] | 0.35 | 0.7 |

[a] Steel that has been hot-dip galvanized according to EN ISO 1461 and with steel composition according to Category A or B of EN ISO 14713-2, Table 1.

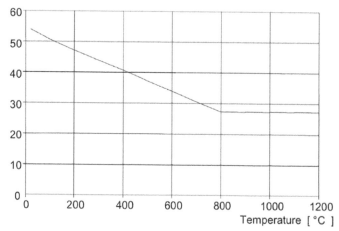

Figure A.1  Variation of thermal conductivity of carbon steel with temperature (CEN, 2005).

temperature. The temperature-dependent values of thermal conductivity for different types of steel are as follows.

### A.1.3.1 Carbon steel (including bolts, welds and fire-resistant steel)

$$20°C \leq \theta_a < 800°C: \quad \lambda_a = 54 - 3.33 \times 10^{-2}\theta_a \left( \frac{W}{mK} \right) \tag{A.1a}$$

$$800°C \leq \theta_a \leq 1200°C: \quad \lambda_a = 27.3 \left( \frac{W}{mK} \right) \tag{A.1b}$$

Figure A.1 plots the temperature-dependent relationship.

### A.1.3.2 Stainless steel

For austenitic and duplex stainless steels:

$$\lambda_a = 14.6 + 1.27 \times 10^{-2} \theta_a \quad \left( \frac{W}{mK} \right) \tag{A.2a}$$

For ferritic stainless steel:

$$\lambda_a = 20.4 + 2.28 \times 10^{-2} \theta_a - 1.54 \times 10^{-5} \theta_a^2 \quad \left( \frac{W}{mK} \right) \tag{A.2b}$$

## A.1.4 Specific heat

Specific heat is the amount of energy needed to raise the temperature of one unit mass of material by one degree in temperature rise. The temperature-dependent values of different types of steel are as follows.

### A.1.4.1 Carbon steel (including bolts, welds and fire-resistant steel)

$$20°C \le \theta_a < 600°C: \quad c_a = 425 + 7.73 \times 10^{-1} \theta_a - 1.69 \times 10^{-3} \theta_a^2$$

$$+ 2.22 \times 10^{-6} \theta_a^3 \quad \frac{J}{kgK} \tag{A.3a}$$

$$600°C \le \theta_a < 735°C: \quad c_a = 666 + \frac{13002}{738 - \theta_a} \left( \frac{J}{kgK} \right) \tag{A.3b}$$

$$735°C \le \theta_a < 900°C: \quad c_a = 545 + \frac{17820}{\theta_a - 731} \left( \frac{J}{kgK} \right) \tag{A.3c}$$

$$900°C \le \theta_a \le 1200°C: \quad c_a = 650 \left( \frac{J}{kgK} \right) \tag{A.3d}$$

These relationships are plotted in Figure A.2.

### A.1.4.2 Stainless steel

Austenitic and duplex stainless steels:

$$c_a = 450 + 0.28 \times \theta_a - 2.91 \times 10^{-4} \theta_a^2 + 1.34 \times 10^{-7} \theta_a^3 \quad \left( \frac{J}{kgK} \right) \tag{A.4a}$$

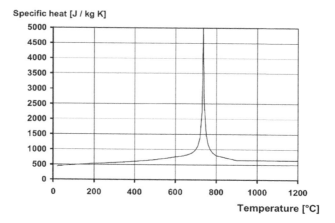

*Figure A.2* Variation of specific heat of carbon steel with temperature (CEN, 2005).

Ferritic stainless steel:

$$c_a = 430 + 0.26 \times \theta_a \left( \frac{J}{\text{kg K}} \right)$$    (A.4b)

## A.1.5 Mechanical properties

The various mechanical properties of steel are expressed as temperature-dependent retention factors, being the ratio of the value of a particular property at elevated temperature to that at ambient temperature.

### A.1.5.1 Normal-strength carbon steel

The general stress-strain curve of carbon steel at elevated temperatures is shown in Figure A.3. It depicts a linear elastic range until reaching the so-called proportional limit stress, followed by an elliptic curve and then a plastic curve until fracture. For S235, S275 and S355 steel, Table A.2 provides the temperature-dependent retention factors for effective yield stress, the proportional stress limit and the Young's modulus.

### A.1.5.2 High-strength steel

Compared to normal-strength carbon steel, there are large uncertainties in mechanical properties of high-strength steel at elevated temperatures. Therefore, extra care must be taken when using high-strength steel in fire. In Europe, EN1993-1-12 (CEN, 2007) recommends that if the following

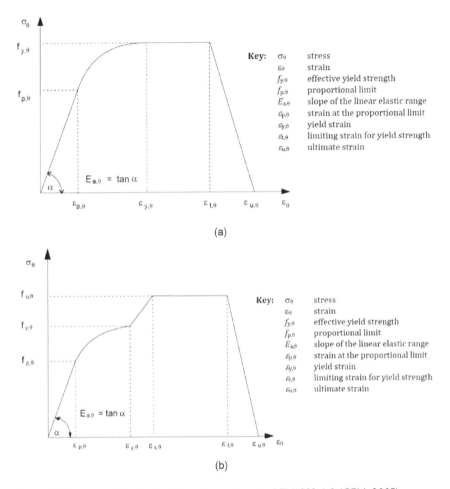

Figure A.3 Stress-strain relationships of a carbon steel EN1993-1-2 (CEN, 2005).

conditions are met, the same temperature-dependent properties in the previous section for normal-strength carbon steel may be used for high-strength steel with steel grades above S460 but not exceeding S690.

$$\frac{f_u}{f_y} \geq 1.05$$

Elongation at failure not less than 10%

$$\varepsilon_u \geq 15\frac{f_y}{E}$$

Table A.2 Reduction factors for carbon steel

| Steel temperature $\theta_a$ | Reduction factors at temperature $\theta_a$ relative to the value of $f_y$ or $E_a$ at 20°C | | |
|---|---|---|---|
| | Reduction factor (relative to $f_y$) for effective yield strength $k_{y,\theta}=f_{y,\theta}/f_y$ | Reduction factor (relative to $f_y$) for proportional limit $k_{p,\theta}=f_{p,\theta}/f_y$ | Reduction factor (relative to $E_a$) for the slope of the linear elastic range $k_{E,\theta}=E_{a,\theta}/E_a$ |
| 20°C | 1.000 | 1.000 | 1.000 |
| 100°C | 1.000 | 1.000 | 1.000 |
| 200°C | 1.000 | 0.807 | 0.900 |
| 300°C | 1.000 | 0.613 | 0.800 |
| 400°C | 1.000 | 0.420 | 0.700 |
| 500°C | 0.780 | 0.360 | 0.600 |
| 600°C | 0.470 | 0.180 | 0.310 |
| 700°C | 0.230 | 0.075 | 0.130 |
| 800°C | 0.110 | 0.050 | 0.090 |
| 900°C | 0.060 | 0.0375 | 0.0675 |
| 1000°C | 0.040 | 0.0250 | 0.0450 |
| 1100°C | 0.020 | 0.0125 | 0.0225 |
| 1200°C | 0.000 | 0.0000 | 0.0000 |

For intermediate values of the steel temperature, linear interpolation may be used.

where

$f_u$ is the ultimate tensile stress,
$f_y$ is the yield stress,
$\varepsilon_u$ is the ultimate strain.

For more detailed information, the reader is referred to specialist publications of Chen et al. (2006), Wang and Lui (2016), Xing et al. (2021), Li and Song (2020), Wu et al. (2021) and Wang et al. (2021).

### A.1.5.3 Stainless steel

Compared to carbon steel, stainless steel can retain higher percentages of strength and stiffness at high temperatures, as shown in Figure A.4.

Table A.3 lists temperature-dependent retention factors for stainless steel.

### A.1.5.4 Bolts and welds

Owing to their different compositions and processing methods (e.g. quenching and tempering for bolts or rapid cooling from molten state for weld) compared to normal steel, the strength retention factors of bolts and

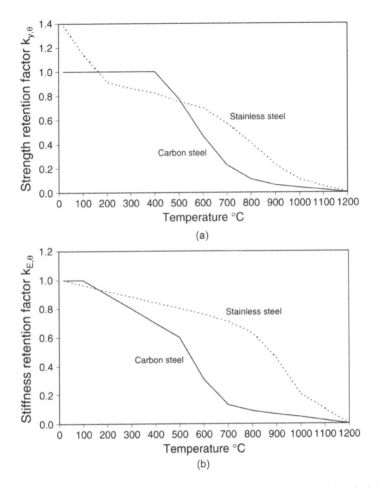

*Figure A.4* Comparison between mechanical properties of stainless steel (grade 1.4301) and carbon steel (Gardner and Baddoo, 2006).

weld are different to those presented previously for normal-strength steel. Table A.4 lists temperature-dependent strength retention factors for bolts and welds.

### A.1.5.5 Fire-resistant steel

FR steel is made by adding a small amount of alloy element Mo. It was originally developed by Sakumoto et al. (1992) to increase the strength and stiffness retention factors of steel at elevated temperatures, in particular in the critical temperature range between about 600°C and 800°C.

*Table A.3* Reduction factors for various types of stainless steel

| Steel temperature $\theta_a$ | Reduction factor (relative to $E_a$) for the slope of the linear elastic range $k_{E.\theta}=E_{a.\theta}/E_a$ | Reduction factor (relative to $f_y$) for proof strength $k_{0.2p.\theta}=f_{0.2p.\theta}/f_y$ | Reduction factor (relative to $f_u$) for tensile strength $k_{u.\theta}=f_{u.\theta}/f_u$ | Factor for determination of the yield strength $f_{y.\theta}$ $k_{2\%.\theta}$ |
|---|---|---|---|---|
| **Grade 1.4301** | | | | |
| 20 | 1.00 | 1.00 | 1.00 | 0.26 |
| 100 | 0.96 | 0.82 | 0.87 | 0.24 |
| 200 | 0.92 | 0.68 | 0.77 | 0.19 |
| 300 | 0.88 | 0.64 | 0.73 | 0.19 |
| 400 | 0.84 | 0.60 | 0.72 | 0.19 |
| 500 | 0.80 | 0.54 | 0.67 | 0.19 |
| 600 | 0.76 | 0.49 | 0.58 | 0.22 |
| 700 | 0.71 | 0.40 | 0.43 | 0.26 |
| 800 | 0.63 | 0.27 | 0.27 | 0.35 |
| 900 | 0.45 | 0.14 | 0.15 | 0.38 |
| 1000 | 0.20 | 0.06 | 0.07 | 0.40 |
| 1100 | 0.10 | 0.03 | 0.03 | 0.40 |
| 1200 | 0.00 | 0.00 | 0.00 | 0.40 |
| **Grade 1.4401/1.4404** | | | | |
| 20 | 1.00 | 1.00 | 1.00 | 0.24 |
| 100 | 0.96 | 0.88 | 0.93 | 0.24 |
| 200 | 0.92 | 0.76 | 0.87 | 0.24 |
| 300 | 0.88 | 0.71 | 0.84 | 0.24 |
| 400 | 0.84 | 0.66 | 0.83 | 0.21 |
| 500 | 0.80 | 0.63 | 0.79 | 0.20 |
| 600 | 0.76 | 0.61 | 0.72 | 0.19 |
| 700 | 0.71 | 0.51 | 0.55 | 0.24 |
| 800 | 0.63 | 0.40 | 0.34 | 0.35 |
| 900 | 0.45 | 0.19 | 0.18 | 0.38 |
| 1000 | 0.20 | 0.10 | 0.09 | 0.40 |
| 1100 | 0.10 | 0.05 | 0.04 | 0.40 |
| 1200 | 0.00 | 0.00 | 0.00 | 0.40 |
| **Grade 1.4571** | | | | |
| 20 | 1.00 | 1.00 | 1.00 | 0.25 |
| 100 | 0.96 | 0.89 | 0.88 | 0.25 |
| 200 | 0.92 | 0.83 | 0.81 | 0.25 |
| 300 | 0.88 | 0.77 | 0.80 | 0.24 |
| 400 | 0.84 | 0.72 | 0.80 | 0.22 |
| 500 | 0.80 | 0.69 | 0.77 | 0.21 |

(Continued)

*Table A.3 (Continued)*  Reduction factors for various types of stainless steel

| Steel temperature $\theta_a$ | Reduction factor (relative to $E_a$) for the slope of the linear elastic range $k_{E,\theta}=E_{a,\theta}/E_a$ | Reduction factor (relative to $f_y$) for proof strength $k_{0.2p,\theta}=f_{0.2p,\theta}/f_y$ | Reduction factor (relative to $f_u$) for tensile strength $k_{u,\theta}=f_{u,\theta}/f_u$ | Factor for determination of the yield strength $f_{y,\theta}$ $k_{2\%,\theta}$ |
|---|---|---|---|---|
| 600 | 0.76 | 0.66 | 0.71 | 0.21 |
| 700 | 0.71 | 0.59 | 0.57 | 0.25 |
| 800 | 0.63 | 0.50 | 0.38 | 0.35 |
| 900 | 0.45 | 0.28 | 0.22 | 0.38 |
| 1000 | 0.20 | 0.15 | 0.11 | 0.40 |
| 1100 | 0.10 | 0.075 | 0.055 | 0.40 |
| 1200 | 0.00 | 0.00 | 0.00 | 0.40 |
| **Grade 1.4003** | | | | |
| 20 | 1.00 | 1.00 | 1.00 | 0.37 |
| 100 | 0.96 | 1.00 | 0.94 | 0.37 |
| 200 | 0.92 | 1.00 | 0.88 | 0.37 |
| 300 | 0.88 | 0.98 | 0.86 | 0.37 |
| 400 | 0.84 | 0.91 | 0.83 | 0.42 |
| 500 | 0.80 | 0.80 | 0.81 | 0.40 |
| 600 | 0.76 | 0.45 | 0.42 | 0.45 |
| 700 | 0.71 | 0.19 | 0.21 | 0.46 |
| 800 | 0.63 | 0.13 | 0.12 | 0.47 |
| 900 | 0.45 | 0.10 | 0.11 | 0.47 |
| 1000 | 0.20 | 0.07 | 0.09 | 0.47 |
| 1100 | 0.10 | 0.035 | 0.045 | 0.47 |
| 1200 | 0.00 | 0.00 | 0.00 | 0.47 |
| **Grade 1.4462** | | | | |
| 20 | 1.00 | 1.00 | 1.00 | 0.35 |
| 100 | 0.96 | 0.91 | 0.93 | 0.35 |
| 200 | 0.92 | 0.80 | 0.85 | 0.32 |
| 300 | 0.88 | 0.75 | 0.83 | 0.30 |
| 400 | 0.84 | 0.72 | 0.82 | 0.28 |
| 500 | 0.80 | 0.65 | 0.71 | 0.30 |
| 600 | 0.76 | 0.56 | 0.57 | 0.33 |
| 700 | 0.71 | 0.37 | 0.38 | 0.40 |
| 800 | 0.63 | 0.26 | 0.29 | 0.41 |
| 900 | 0.45 | 0.10 | 0.12 | 0.45 |
| 1000 | 0.20 | 0.03 | 0.04 | 0.47 |
| 1100 | 0.10 | 0.015 | 0.02 | 0.47 |
| 1200 | 0.00 | 0.00 | 0.00 | 0.47 |

*Table A.4* Strength reduction factors for bolts and welds

| Temperature $\theta_a$ | For bolts. $k_{b,\theta}$ (tension and shear) | For welds. $k_{w,\theta}$ |
|---|---|---|
| 20 | 1.000 | 1.000 |
| 100 | 0.968 | 1.000 |
| 150 | 0.952 | 1.000 |
| 200 | 0.935 | 1.000 |
| 300 | 0.903 | 1.000 |
| 400 | 0.775 | 0.876 |
| 500 | 0.550 | 0.627 |
| 600 | 0.220 | 0.378 |
| 700 | 0.100 | 0.130 |
| 800 | 0.067 | 0.074 |
| 900 | 0.033 | 0.018 |
| 1000 | 0.000 | 0.000 |

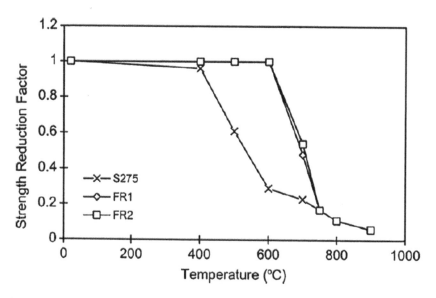

*Figure A.5* Strength reduction factors for the fire-resistant steels and carbon steel.

According to Kelly and Sha (1999), the yield stress retention factors of FR steel is as shown in Figure A.5. For FR bolts as reported by Sakumoto et al. (1993), the yield stress and Young's modulus retention factors, as well as elongation of FR steel are as presented the Table A.5.

More recent work by Kumar et al. (2021) has expanded the database on FR steel as presented in Table A.6.

Table A.5 Strength reduction factors for fire-resistant (FR) bolts (Sakumoto et al., 1993)

| Temperature (°C) | Reduction factors | | Young's modulus FR bolts | Elongation% FR bolts |
|---|---|---|---|---|
| | Tensile strength | | | |
| | Carbon steel bolts | FR bolts | | |
| 20 | 1 | 1 | 1 | 14 |
| 300 | 0.903 | 0.959 | 0.898 | 16 |
| 400 | 0.775 | 0.874 | 0.896 | 13 |
| 500 | 0.55 | 0.747 | 0.79 | 13 |
| 550 | 0.385 | 0.624 | 0.716 | 17 |
| 600 | 0.22 | 0.43 | 0.608 | 23 |
| 650 | 0.165 | 0.273 | 0.444 | 34 |
| 700 | 0.1 | 0.166 | 0.333 | 57 |
| 800 | 0.067 | 0.074 | 0.234 | – |
| 900 | 0.033 | 0.033 | – | – |
| 1000 | 0 | 0 | – | – |

Table A.6 Strength and stiffness retention factors for different types of fire-resistant steels (Kumar et al., 2021)

**Fire-resistant steel with 0.126% of alloying element Mo**

| Yield strength | $k_{y,\theta} = 3 \times 10^{-12}\theta^4 - 7 \times 10^{-9}\,\theta^3 + 2 \times 10^{-6}\,\theta^2 - 5 \times 10^{-4}\,\theta + 1.0137$ | $20°C \leq \theta \leq 800°C$ |
|---|---|---|
| Elastic modulus | $k_{E,\theta} = 2 \times 10^{-6}\,\theta^2 - 9 \times 10^{-4}\,\theta + 1.0179$ | $20°C \leq \theta \leq 200°C$ |
| | $= -7 \times 10^{-9}\,\theta^3 + 9 \times 10^{-6}\,\theta^2 - 4.5 \times 10^{-3}\theta + 1.519$ | $200°C \leq \theta \leq 500°C$ |
| | $= -9 \times 10^{-9}\,\theta^3 + 2 \times 10^{-5}\,\theta^2 - 0.0195\theta + 5.864$ | $500°C \leq \theta \leq 800°C$ |

**Fire-resistant steel with 0.1% of alloying element Mo**

| Yield strength | $k_{y,\theta} = 4 \times 10^{-12}\,\theta^4 - 8 \times 10^{-9}\,\theta^3 + 4 \times 10^{-6}\,\theta^2 - 1.1 \times 10^{-3}\,\theta + 1.0233$ | $20°C \leq \theta \leq 800°C$ |
|---|---|---|
| Elastic modulus | $k_{E,\theta} = 8 \times 10^{-7}\,\theta^2 - 3 \times 10^{-4}\,\theta + 1.0064$ | $20°C \leq \theta \leq 200°C$ |
| | $= -8 \times 10^{-9}\,\theta^3 + 1 \times 10^{-5}\,\theta^2 - 5.1 \times 10^{-3}\,\theta + 1.65$ | $200°C \leq \theta \leq 500°C$ |
| | $= 2 \times 10^{-8}\,\theta^3 - 3 \times 10^{-5}\,\theta^2 + 0.0113\theta + 0.545$ | $500°C \leq \theta \leq 800°C$ |

# REFERENCES

CEN 2002. EN 1991-1-2 Eurocode 1: Actions on structures–Part 1–2: General actions– Action on structures exposed to fire. *British Standards Institution.*

CEN 2005. EN 1993-1-2 Eurocode 3: Design of Steel Structures–Part 1–2: General Rules–Structural Fire Design. *British Standards Institution.*

CEN 2007. EN 1993-1-12 Eurocode 3: Design of Steel Structures–Part 1–12: Additional rules for the extension of EN 1993 up to steel grades S700. *British Standards Institution.*

Chen, J., Young, B. & Uy, B. 2006. Behavior of high strength structural steel at elevated temperatures. *Journal of Structural Engineering*, 132, 1948–1954.

Gardner, L. & Baddoo, N. R. 2006. Fire testing and design of stainless steel structures. *Journal of Constructional Steel Research*, 62, 532–543.

Kelly, F. S. & Sha, W. 1999. A comparison of the mechanical properties of fire-resistant and S275 structural steels. *Journal of Constructional Steel Research*, 50, 223–233.

Kumar, W., Sharma, U. K. & Shome, M. 2021. Mechanical properties of conventional structural steel and fire-resistant steel at elevated temperatures. *Journal of Constructional Steel Research*, 181, 106615.

Li, G.-Q. & Song, L.-X. 2020. Mechanical properties of TMCP Q690 high strength structural steel at elevated temperatures. *Fire Safety Journal*, 116, 103190.

Sakumoto, Y., Keira, K., Furumura, F. & Ave, T. 1993. Tests of fire-resistant bolts and joints. *Journal of Structural Engineering*, 119, 3131–3150.

Sakumoto, Y., Yamaguchi, T., Ohashi, M. & Saito, H. 1992. High-temperature properties of fire-resistant steel for buildings. *Journal of Structural Engineering*, 118, 392–407.

Wang, F. & Lui, E. 2016. Behavior of high strength steels under and after high temperature exposure: A review. *Journal of Steel Structures Construction*, 2, 10.

Wang, W., Li, X. & Al-Azzani, H. 2021. Experimental study on local buckling of high-strength Q960 steel columns at elevated temperatures. *Journal of Constructional Steel Research*, 183, 106716.

Wu, Y., Fan, S., He, B., Li, C. & Zhou, H. 2021. Experimental study and numerical simulation analysis of the fire resistance of high-strength steel columns with rectangular sections under axial compression. *Fire Safety Journal*, 121, 103266.

Xing, Y., Wang, W. & Al-Azzani, H. 2021. Assessment of thermal properties of various types of high-strength steels at elevated temperatures. *Fire Safety Journal*, 122, 103348.

# Appendix B: Examples of Checking for shear and bending of connections

## B.1 SHEAR RESISTANCE OF A FIN PLATE CONNECTION

Figure B.1 presents details of the connection. The critical temperature of the connected beam is assumed to be 543.83°C.

The material properties of the connection are as follows:

- M20 Grade 8.8 bolts:
  $A_s = 245 mm^2$
  $d = 20 mm$
  $f_{yb} = 640 N/mm^2$
  $f_{ub} = 800 N/mm^2$
- Fillet welds:
  Leg length: 8 mm
  Throat thickness a = 5.7 mm
  $\alpha_v = 0.6$ (constant for bolt shear resistance)

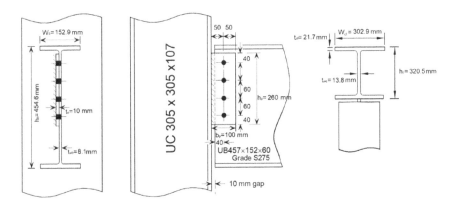

*Figure B.1* Details of fin plate connection.

- S275 steel:
    Yield strength $f_y = 275$ N/mm²
    Ultimate strength $f_u = 430$ N/mm²
    Young's modulus $E = 205000$ N/mm²

## Beam reaction force:
$V_{Ed} = 124.163$ kN
At the critical temperature of 543.83°C, the various retention factors are:
  Bolt: 0.4 (see Annex A.1.4.4)
  Weld: 0.52 (see Annex A.1.4.4)
  Steel: 0.64 (see Annex A.1.4.1)
Note: since equations for the following detailed checks are not given in this book, reference is made to clause numbers of EN 1993-1-8 (CEN, 2005).

## Bolt – shear resistance (Clause 3.6 Table 3.4)
Shear capacity of each bolt: $F_{v,\,Rd} = \alpha_v k_{b,\,\theta} f_{ub} A_s = 0.6 \times 0.373 \times 800 \times 245 = 51.71$ kN

Total shear resistance of the bolt group $= nF_{v,\,Rd,\,fi} (1 + 6z/(n+1) p_1)^{0.5} = 64 \times 51.7/\sqrt{(1 + (6 \times 50/[(4+1) \times 60])^2)} = 146.26$ kN $> V_{Ed,\,fi} = 124.16$ kN, ∴ so ok.

## Fin plate – bearing resistance (Clause 3.6 Table 3.4)
Edge bolt in vertical direction: $F_{b,\,fi,\,Rd} = k_1 \alpha_b k_{y,\,\theta} f_u dt = 2.5 \times 0.61 \times 0.64 \times 275 \times 20 \times 10 = 53.68$ kN

Inner bolt in vertical direction: $F_{b,\,fi,\,Rd} = k_1 \alpha_b k_{y,\,\theta} f_u dt = 2.5 \times 0.65 \times 0.64 \times 275 \times 20 \times 10 = 57.2$ kN

Edge bolt in horizontal direction $F_{b,\,Rd} = k_1 \alpha_b k_{y,\,\theta} f_u dt = 2.5 \times 0.75 \times 0.64 \times 275 \times 20 \times 10 = 66.0$ kN

$\beta = 6 \times 50/(60 \times 4 \times (4+1)) = 0.25$

Overall plate bearing resistance for the bolt group $V_{b,\,Rd} = 1/([(1/n) + \alpha/F_{b,\,fi,\,Rd}]^2 + (\beta/F_{b,\,Rd})^2] = 1/\sqrt{[(1/4/53.68)^2 + (0.25/66.0)^2]} = 166.58$ kN $> V_{Ed} = 124.16$ kN ∴ so ok.

## Fin plate – shear resistance (Clause 3.10.2)
Of the gross cross-section $= A_v k_{y,\,\theta} fy/(1.27\sqrt{3}) = (10 \times 260 \times 0.64 \times 275)/(1.27 \times \sqrt{3} \times 1.0 \times 1000) = 208.03$ kN

Of the net cross-section $= A_{v,\,net} k_{y,\,\theta} f_u/\sqrt{3} = (10 \times (260 - 4 \times 22) \times 0.64 \times 275/(\sqrt{3} \times 1.0 \times 1000) = 174.78$ kN

Block shear resistance $= 0.5 k_{y,\,\theta} f_u A_{nt} + k_{y,\,\theta} f_y A_{nv}/\sqrt{3} = 0.001 \times [0.5 \times 0.64 \times 275 \times (10 \times (50-22 \times 0.5))/1.0] + 0.001 \times [0.64 \times 275 \times 10 \times (260 - 40 - (4 - 0.5) \times 22)/\sqrt{3} \times 1.0] = 179.63$ kN

Min (Gross shear resistance, Net shear resistance, Block shear resistance) $= 174.78$ kN $> V_{Ed,\,fi} = 124.16$ kN, so ok

## Fin plate – lateral tortional buckling resistance

$z < t/0.15 = 50$ mm $< (10/0.15)$. So, short fin plate

$(W_{el}/z)\, k_{y,\theta} f_y = ([(10 \times 260^2)/6]/50) \times 0.64 \times 275/1.0 \times 1000 = 396.59$ kN, so ok

## Beam web – bearing resistance (Clause 3.6 Table 3.4)

Edge bolt in the vertical direction: $F_{b,Rd} = k_1 \alpha_b k_{y,\theta} f_u dt = 2.5 \times 1.0 \times 0.64 \times 275 \times 20 \times 8.1 = 71.28$ kN

Inner bolt in the vertical direction: $F_{b,Rd} = k_1 \alpha_b k_{y,\theta} f_u dt = 2.5 \times 0.66 \times 0.64 \times 275 \times 20 \times 8.1 = 47.04$ kN

In the horizontal direction $F_{b,Rd} = k_1 \alpha_b k_{y,\theta} f_u dt = 2.5 \times 0.61 \times 0.64 \times 275 \times 20 \times 8.1 = 43.48$ kN

Overall beam web bearing resistance for the bolt group $V_{b,Rd} = 1/([(1/n) + \alpha/F_{b,fi,Rd}]^2 + (\beta/F_{b,Rd})^2] = 1/\sqrt{[(1/4/45.48)^2 + (0.25/43.48)^2]} = 127.72$ kN $> V_{Ed,fi} = 124.16$ kN, so ok.

## Beam web – shear resistance (Clause 3.10.2)

Of the gross cross-section $= A_v k_{y,\theta} f_y/(1.27\sqrt{3}) = 3682.26 \times 0.64 \times 275/(\sqrt{3} \times 1.0 \times 1000) = 374.17$ kN

OF the net cross-section $= A_{v,net} k_{y,\theta} f_u/\sqrt{3} = (3682.26 - 4 \times 22 \times 8.1) \times 0.64 \times 275/(\sqrt{3} \times 1.0 \times 1000) = 301.74$ kN

Block shear resistance $= 0.5\, k_{y,\theta} f_u A_{nt} + k_{y,\theta} f_y A_{nv}/\sqrt{3} = 0.001 \times [0.5 \times 0.64 \times 275 \times (8.1 \times (40 - 22 \times 0.5))/1.0] + 0.001 \times [0.64 \times 275 \times (8.1 \times (90 + (4 - 1) \times 60 - (4 - 0.5) \times 22)/\sqrt{3} \times 1.0] = 179.52$ kN

Min (Gross shear resistance, Net shear resistance, Block shear resistance) $= 179.52$ kN $> V_{Ed,fi} = 124.16$ kN, so ok

## Weld shear resistance (Clause 4.5.3.3)

$(k_{w,\theta} l(a_w/\sqrt{2})(f_y/\sqrt{3})/\beta_w = 0.52 \times 2 \times 260 \times 5.7 \times (275/\sqrt{3})/0.85 = 287.9 > 124.16$ kN, so ok.

## Summary of results

Bolts in shear: 146.26 kN

Fin plate in bearing: 166.58 kN

Fin plate in shear: 174.78 kN

Fin plate lateral torsional buckling: 396.59 kN

Beam web in bearing: 127.72 kN

Beam web in shear: 179.52 kN

Weld in shear: 287.9 kN

## B.2 SHEAR AND BENDING RESISTANCES OF AN EXTENDED ENDPLATE CONNECTION

Refer to the numerical example in Chapter 5 for detailed connection arrangement and geometrical information (Figure 5.7b). The temperature of the connection is 584.7°C, which is the same as the critical temperature of the connected beam, taken from the second example in Chapter 2.

**Summary of beam reaction forces:**

Shear force: $V_{Ed, fi} = R_{fi}/2 = 324.5/2 = 162.25$ kN

Bending moment = 150.57kN·m (assumed to be 50% of the plastic bending moment of the connected beam for calculation of the beam critical temperature)

At the beam critical temperature of 584.7°C, reduction factors of the various components are as follows:

Yield and ultimate stress reduction factor for bolt: $k_{b, \theta} = 0.2705$ (see Annex A.1.4.4).

Yield stress reduction factor for weld: $k_{b, \theta} = 0.42$ (see Annex A.1.4.4)

Yield stress reduction factor for steel: $k_{y, \theta} = 0.5174$ (see Annex A.1.4.1)

$\alpha_v = 0.6$ (constant for bolt shear resistance)

**Shear resistance check:**

Shear resistance of M24 Grade 8.8 bolt in single shear $= \alpha_v k_{b, \theta} f_{ub} A_s = 0.6 \times 0.2705 \times 800 \times 353/1.0 = 45.83$ kN

Bearing resistance of a single M24 Grade 8.8 bolt in the thinnest plate (i.e. 20.5 mm column flange) $= k_{y, \theta} k_1 \alpha_b f_u d t_p = 0.52 \times 0.61 \times 2.5 \times 265 \times 24 \times 20.5/1000 = 103.39$kN

Plate bearing resistance > bolt shear resistance, therefore bolt shear failure governs.

The tension bolts are assumed to be fully loaded in tension. Therefore, their shear resistance is 28% of their resistance without tension (from Table 3.4 of EN1993-1-8 (CEN, 2005)).

Total bolt shear resistance $= (2 + 6 \times 0.28) \times 45.83 = 168.65$ kN $> 162.25$ kN, so ok.

**Weld in shear:**

$k_{w, \theta} l (a_w/\sqrt{2}) (f_y/\sqrt{3})/\beta_w = 0.42 \times 2 \times 533 \times (8/\sqrt{2}) \times (275/\sqrt{3})/(0.85 \times 1000) = 473$kN $> 162.25$ kN, so ok.

Checking for bending resistance:

| Row 1 | Resistance | Row 1 |
|---|---|---|
| | **Column flange in bending** | |

**Mode 1 (4 plastic hinges)**

$F_{T,1,Rd} = (8n - 2e_w) M_{pl,fi,Rd}/(2 mn - e_w (m + n))$

$M_{pl,fi,Rd} = 0.25 k_y \Sigma l_{eff,1} t_f^2 f_y /\gamma_{fi} = 0.25 \times 0.5174 \times 210 \times 20.5^2 \times 265/1.0$
$= 3025.09$ kN·mm

$e_w = d_w / 4 = 39.55/4 = 9.9$ mm

$F_{T,1,Rd} = ((8 \times 41.8 - 2 \times 9.9) \times 3025.09)/(2 \times 33.4 \times 41.8 - 9.9 \times (33.4 + 41.8)) = 464.74$ kN

**Mode 2 (2 plastic hinges and bolt failure)**

$F_{T,2,Rd} = (2 M_{pl,fi,Rd} + n \Sigma F_{t,b,Rd})/(m + n)$

$M_{pl,fi,Rd} = 0.25 k_y \Sigma l_{eff,2} t_f^2 f_y /\gamma_{fi} = 0.25 \times 0.5174 \times 233 \times 20.5^2 \times 265/1.0$
$= 3356.42$ kN·mm

$F_{t,b,Rd} = (k_2 k_{yb} f_{ub} A_s)/\gamma_{fi} = 0.9 \times 0.2705 \times 800 \times 353/1.0 = 68.75$ kN

$F_{T,2,Rd} = (2 \times 3356.42 + 41.48 \times (2 \times 68.75))/(33.4 + 41.8) = 164.23$ kN

**Mode 3 (bolt failure)**

$F_{T,3,Rd} = \Sigma F_{t,b,Rd} = 2 \times (0.9 \times 0.2705 \times 800 \times 353/1.0) = 68.75$ kN

**Resistance = Min (Mode 1, Mode 2, Mode 3) = 137.5kN**

**Column web in transverse tension**

$F_{t,wc,Rd} = \omega b_{eff,t,wc} t_{wc} k_y f_{yl} /\gamma_{fi} = 1.0 \times 233 \times 12.8 \times 0.5174 \times 265/1000$
$= 408.92$ kN

**Endplate in bending**

**Mode I (4 plastic hinges)**
$F_{T,1,Rd} = (8n - 2e_w) M_{pl,fi,Rd}/(2mn - e_w(m+n))$
$M_{pl,fi,Rd} = 0.25 k_y \Sigma l_{eff,1} t_p^2 f_y/\gamma_{fi} = 0.25 \times 0.5174 \times 125 \times 25^2 \times 265/1.0$
$= 2677.95$ kN·mm

$e_w = d_w/4 = 39.55/4 = 9.9$mm
$F_{T,1,Rd} = ((8 \times 38 - 2 \times 9.9) \times 2677.95)/(2 \times 30.4 \times 38 - 9.9 \times (30.4+38)) = 465.99$ kN

**Mode 2 (2 plastic hinges and bolt failure)**
$F_{T,2,Rd} = (2 M_{pl,fi,Rd} + n \Sigma F_{t,b,Rd})/(m+n)$
$M_{pl,fi,Rd} = 0.25 k_y \Sigma l_{eff,2} t_p^2 f_y/\gamma_{fi} = 0.25 \times 0.5174 \times 125 \times 25^2 \times 265/1.0$
$= 2677.95$ kN·mm

$F_{t,b,Rd} = (k_2 k_{yb} f_{ub} A_s)/\gamma_{fi} = 0.9 \times 0.2705 \times 800 \times 353/1.0 = 68.75$ kN
$F_{T,2,Rd} = (2 \times 2677.95 + 38 \times (2 \times 68.75))/(30.4+38) = 154.69$ kN

**Mode 3 = 94.54 kN (bolt failure)**
$F_{T,3,Rd} = 2 \times (0.9 \times 0.2705 \times 800 \times 353/1.0) = 137.5$ kN

**Resistance = Min (Mode I, Mode 2, Mode 3) = 137.5 kN**

Row 2    **Column flange in bending**
**Mode I (4 plastic hinges)**
$F_{T,1,Rd} = (8n - 2e_w) M_{pl,fi,Rd}/(2mn - e_w(m+n))$
$M_{pl,fi,Rd} = 0.25 k_y \Sigma l_{eff,1} t_p^2 f_y/\gamma_{fi} = 0.25 \times 0.5174 \times 210 \times 20.5^2 \times 265/1.0$
$= 3025.09$ kN·mm

$e_w = d_w/4 = 39.55/4 = 9.9$mm
$F_{T,1,Rd} = ((8 \times 41.8 - 2 \times 9.9) \times 3025.09)/(2 \times 33.4 \times 41.8 - 9.9 \times (33.4+41.8)) = 464.75$ kN

**Mode 2 (2 plastic hinges and bolt failure)**
$F_{T,2,Rd} = (2 M_{pl,fi,Rd} + n \Sigma Ft_{t,b,Rd})/(m+n)$
$M_{pl,fi,Rd} = 0.25 k_y \Sigma l_{eff,2} t_p^2 f_y/\gamma_{fi} = 0.25 \times 0.5174 \times 233 \times 20.5^2 \times 265/1.0$
$= 3356.42$ kN·mm
$F_{t,b,Rd} = (k_2 k_{yb} f_{ub} A_s)/\gamma_{fi} = 0.9 \times 0.2705 \times 800 \times 353/1.0 = 68.75$ kN
$F_{T,2,Rd} = (2 \times 3356.42 + 41.48 \times (2 \times 68.75))/(33.4+41.8) = 164.23$ kN

Row 2

## Mode 3 (bolt failure)

$F_{T,3,Rd} = \Sigma F_{t,b,Rd} = 2 \times (0.9 \times 0.2705 \times 800 \times 353/1.0) = 137.5$ kN

## Resistance = Min (Mode 1, Mode 2, Mode 3) = 137.5 kN

### Column web in transverse tension

$F_{t,wc,Rd} = \omega b_{eff,t,wc} t_{wc} k_y f_y / \gamma_{fi} = 1.0 \times 233 \times 12.8 \times 0.5174 \times 265/1000$
$= 408.92$ kN

## End plate in bending

### Mode 1 (4 plastic hinges)

$F_{T,1,Rd} = (8n - 2e_w) M_{pl,fi,Rd} / (2 mn - e_w (m+n))$

$M_{pl,fi,Rd} = 0.25 k_y \Sigma l_{eff,1} t_p^2 f_y / \gamma_{fi} = 0.25 \times 0.5174 \times 243 \times 25^2 \times 265/1.0$
$= 5205.93$ kN·mm

$e_w = d_w/4 = 39.55/4 = 9.9$ mm

$F_{T,1,Rd} = ((8 \times 48.3 - 2 \times 9.9) \times 5205.93)/(2 \times 38.6 \times 48.3 - 9.9 \times (38.6+48.3)) = 665.33$ kN

## Mode 2 (2 plastic hinges and bolt failure)

$F_{T,2,Rd} = (2 M_{pl,fi,Rd} + n\Sigma F_{t,b,Rd})/(m+n)$

$M_{pl,fi,Rd} = 0.25 k_y \Sigma l_{eff,2} t_p^2 f_y / \gamma_{fi} = 0.25 \times 0.5174 \times 290 \times 25^2 \times 265/1.0$
$= 6212.84$ kN·mm

$F_{t,b,Rd} = (k_2 k_{yb} f_{ub} A_s)/\gamma_{fi} = 0.9 \times 0.2705 \times 800 \times 353/1.0 = 68.75$ kN

$F_{T,2,Rd} = (2 \times 6212.84 + 48.3 \times (2 \times 68.75))/(38.6+48.3) = 219.41$ kN

## Mode 3 (bolt failure)

$F_{T,3,Rd} = \Sigma F_{t,b,Rd} = 2 \times (0.9 \times 0.2705 \times 800 \times 353/1.0) = 137.5$ kN

## Resistance = Min (Mode 1, Mode 2, Mode 3) = 137.5 kN

### Beam web in tension

$F_{t,wb,Rd} = \omega b_{eff,t,wc} t_{wb} k_y f_y / \gamma_{fi} = 1.0 \times 243 \times 10.1 \times 0.5174 \times 275/1000$
$= 349.21$ kN

## Row 1 and 2 combined – column flange in bending

### Column flange in bending

#### Mode 1 (4 plastic hinges)

$F_{T,1,Rd} = (8n - 2e_w) M_{pl,fi,Rd} / (2 mn - e_w (m+n))$

$M_{pl,fi,Rd} = 0.25 \ k_y \ \Sigma l_{eff,1} \ t_f^2 \ f_y / \gamma_{fi} = 0.25 \times 0.5174 \times 332 \times 20.52 \times 265/1.0 = 4782.53 \ \text{kN·mm}$

$e_w = d_w / 4 = 39.55/4 = 9.9 \ \text{mm}$

$F_{T,1,Rd} = ((8 \times 41.8 - 2 \times 9.9) \times 4782.53)/(2 \times 33.4 \times 41.8 - 9.9 \times (33.4+41.8)) = 734.75 \ \text{kN}$

#### Mode 2 (2 plastic hinges and bolt failure)

$F_{T,2,Rd} = (2 \ M_{pl,fi,Rd} + n \ \Sigma F_{t,b,Rd})/(m+n)$

$M_{pl,fi,Rd} = 0.25 \ k_y \ \Sigma l_{eff,2} \ t_f^2$

$f_y / \gamma_{fi} = 0.25 \times 0.5174 \times 332 \times 20.5^2 \times 265/1.0 = 4782.53 \ \text{kN·mm}$

$F_{t,b,Rd} = (k_2 \ k_{y,b} \ f_{ub} \ A_s)/\gamma_{fi} = 0.9 \times 0.2705 \times 800 \times 353/1.0 = 68.75 \ \text{kN}$

$F_{T,2,Rd} = (2 \times 4782.53 + 41.8 + (4 \times 68.75))/(33.4+41.8)$
$= 280.05 \ \text{kN}$

#### Mode 3 (bolt failure)

$F_{T,3,Rd} = \Sigma F_{t,b,Rd} = 4 \times (0.9 \times 0.2705 \times 800 \times 353/1.0) = 275 \ \text{kN}$

**Resistance = Min (Mode 1, Mode 2, Mode 3) = 275 kN**

Row 1 and 2 combined – column web in transverse tension

$F_{t,wc,Rd} = \omega b_{eff,t,wc} \ t_{wc} \ k_y f_y / \gamma_{fi} = 1.0 \times 332 \times 12.8 \times 0.5174 \times 265/1000$
$= 582.67 \ \text{kN}$

**Therefore, the resistance of row 2 bolt row 2 on the column side is = Min (275, 582.67) – 137.5 = 137.5 kN**

Row 3

## Column flange in bending

#### Mode 1 (4 plastic hinges)

$F_{T,1,Rd} = (8n - 2e_w) M_{pl,fi,Rd} / (2 mn - e_w (m+n))$

$M_{pl,fi,Rd} = 0.25 \ k_y \ \Sigma l_{eff,1} \ t_f^2 \ f_y / \gamma_{fi} = 0.25 \times 0.5174 \times 210 \times 20.5^2 \times 265/1.0$
$= 3025.09 \ \text{kN·mm}$

$e_w = d_w / 4 = 39.55/4 = 9.9 \ \text{mm}$

$F_{T,1,Rd} = ((8 \times 41.8 - 2 \times 9.9) \times 3025.09)/(2 \times 33.4 \times 41.8 - 9.9 \times (33.4+41.8)) = 464.75 \ \text{kN}$

Row 3

### Mode 2 (2 plastic hinges and bolt failure)

$F_{T,2,Rd} = (2 M_{pl,fi,Rd} + n \Sigma F_{t,b,Rd})/(m+n)$

$M_{pl,fi,Rd} = 0.25 k_y \Sigma l_{eff} 2 t_f^2 f_y/\gamma_{fi} = 0.25 \times 0.5174 \times 233 \times 20.5^2 \times 265/1.0 = 3356.42$ kN·mm

$F_{t,b,Rd} = (k_2 k_{y,b} f_{ub} A_s)/\gamma_{fi} = 0.9 \times 0.2705 \times 800 \times 353/1.0 = 68.75$ kN

$F_{T,2,Rd} = (2 \times 3356.42 + 41.48 \times (2 \times 68.75))/(33.4 + 41.8) = 165.11$ kN

### Mode 3 (bolt failure)

$F_{T,3,Rd} = \Sigma F_{t,b,Rd} = 2 \times (0.9 \times 0.2705 \times 800 \times 353/1.0) = 137.5$ kN

### Resistance = Min (Mode 1, Mode 2, Mode 3) = 137.5 kN

### Column web in transverse tension

$F_{t,wc,Rd} = \omega b_{eff,t,wc} t_{wc} k_y f_y/\gamma_{fi} = 1.0 \times 233 \times 12.8 \times 0.5174 \times 265/1000 = 408.92$ kN
End plate in bending

### Mode 1 (4 plastic hinges)

$F_{T,1,Rd} = (8n - 2e_w) M_{pl,fi,Rd}/(2mn - e_w (m+n))$

$M_{pl,fi,Rd} = 0.25 k_y \Sigma l_{eff,1} t_p^2 f_y/\gamma_{fi} = 0.25 \times 0.5174 \times 243 \times 25^2 \times 265/1.0 = 5205.93$ kN·mm

$e_w = d_w/4 = 39.55/4 = 9.9$ mm

$F_{T,1,Rd} = ((8 \times 48.3 - 2 \times 9.9) \times 5205.93)/(2 \times 38.6 \times 48.3 - 9.9 \times (38.6 + 48.3)) = 665.33$ kN

### Mode 2 = (2 plastic hinges and bolt failure)

$F_{T,2,Rd} = (2 M_{pl,fi,Rd} + n \Sigma F_{t,b,Rd})/(m+n)$

$M_{pl,fi,Rd} = 0.25 k_y \Sigma l_{eff2} t_p^2 f_y/\gamma_{fi} = 0.25 \times 0.5174 \times 248 \times 25^2 \times 265/1.0 = 5313.05$ kN·mm

$F_{t,b,Rd} = (k_2 k_{y,b} f_{ub} A_s)/\gamma_{fi} = 0.9 \times 0.2705 \times 800 \times 353/1.0 = 68.75$ kN

$F_{T,2,Rd} = (2 \times 5313.05 + 48.3 \times (2 \times 68.75))/(38.6 + 48.3) = 198.7$ kN

### Mode 3 (bolt failure)

$\Sigma F_{t,b,Rd} = 2 \times (0.9 \times 0.2705 \times 800 \times 353/1.0) = 137.5$ kN

### Resistance = Min (Mode 1, Mode 2, Mode 3) = 137.5 kN

### Beam web in tension

$F_{t,wb,Rd} = \omega b_{eff,t,wc} t_{wb} k_y f_y/\gamma_{fi} = 1.0 \times 243 \times 10.1 \times 0.5174 \times 275/1000 = 349.21$ kN

### Row 1, 2 and 3 combined – column flange in bending

#### Column flange in bending

#### Mode 1 (4 plastic hinges)

$F_{T,1,Rd} = (8n - 2e_w) \, M_{pl,fi,Rd} / (2 \, mn - e_w \, (m+n))$

$M_{pl,fi,Rd} = 0.25 \, k_y \, \Sigma l_{eff,1} \, t_f^2 \, f_y / \gamma_{fi} = 0.25 \times 0.5174 \times 422 \times 20.5^2 \times 265/1.0$
$= 6079 \text{ kN·mm}$

$e_w = d_w/4 = 39.55/4 = 9.9 \text{ mm}$

$F_{T,1,Rd} = ((8 \times 41.8 - 2 \times 9.9) \times 6079)/(2 \times 33.4 \times 41.8 - 9.9 \times (33.4 + 41.8)) = 933.93 \text{ kN}$

#### Mode 2 (2 plastic hinges and bolt failure)

$F_{T,2,Rd} = (2 \, M_{pl,fi,Rd} + n \, \Sigma F_{t,b,Rd})/(m+n)$

$M_{pl,fi,Rd} = 0.25 \, k_y \, \Sigma l_{eff,2} \, t_f^2 \, f_y / \gamma_{fi} = 0.25 \times 0.5174 \times 422 \times 20.5^2 \times 265/1.0 = 6079 \text{ kN·mm}$

$F_{t,b,Rd} = (k_2 \, k_{yb} \, f_{ub} \, A_s)/\gamma_{fi} = 0.9 \times 0.2705 \times 800 \times 353/1.0 = 68.75 \text{ kN}$

$F_{T,2,Rd} = (2 \times 6079 + 41.8 \times (6 \times 68.75))/(33.4 + 41.8) = 390.29 \text{ kN}$

#### Mode 3 (bolt failure)

$F_{T,3,Rd} = \Sigma F_{t,b,Rd} = 6 \times (0.9 \times 0.2705 \times 800 \times 353/1.0) = 412.5 \text{ kN}$

**Resistance = Min (Mode 1, Mode 2, Mode 3) = 390.29 kN**

Row 1, 2 and 3 combined – column web in transverse tension

$F_{t,wc,Rd} = \omega \, b_{eff,t,wc} \, t_{wc} \, k_y \, f_y / \gamma_{fi} = 1.0 \times 422 \times 12.8 \times 0.5174 \times 265/1000 = 740.62 \text{ kN}$

**Therefore, the resistance of row 3 bolt on the column side is = Min (390.29, 740.62) − 275 = 115.29 kN**

### Row 2 and 3 combined – column flange in bending

#### Mode 1 (4 plastic hinges)

$F_{T,1,Rd} = (8n - 2e_w) \, M_{pl,fi,Rd} / (2mn - e_w \, (m+n))$

$M_{pl,fi,Rd} = 0.25 \, k_y \, \Sigma l_{eff,1} \, t_f^2 \, f_y / \gamma_{fi} = 0.25 \times 0.5174 \times 323 \times 20.5^2 \times 265/1.0 = 4652.89 \text{ kN·mm}$

$e_w = d_w/4 = 39.55/4 = 9.9 \text{ mm}$

$F_{T,1,Rd} = ((8 \times 41.8 - 2 \times 9.9) \times 4652.89)/(2 \times 33.4 \times 41.8 - 9.9 \times (33.4 + 41.8)) = 714.83 \text{ kN}$

**Mode 2 (2 plastic hinges and bolt failure)**

$F_{T,2,Rd} = (2 M_{pl,fi,Rd} + n \Sigma F_{t,b,Rd})/(m+n)$

$M_{pl,fi,Rd} = 0.25 k_y \Sigma l_{eff,2} t_f^2 f_y/\gamma_{fi} = 0.25 \times 0.5174 \times 323 \times 20.5^2 \times 265/1.0 = 4652.89$ kN·mm

$F_{t,b,Rd} = (k_2 k_{y,b} f_{ub} A_s)/\gamma_{fi} = 0.9 \times 0.2705 \times 800 \times 353/1.0 = 68.75$ kN

$F_{T,2,Rd} = (2 \times 4652.89 + 41.8 \times (4 \times 68.75))/(33.4 + 41.8) = 276.61$ kN

**Mode 3 (bolt failure)**

$F_{T,3,Rd} = \Sigma F_{t,b,Rd} = 4 \times (0.9 \times 0.2705 \times 800 \times 353/1.0) = 274.99$ kN

**Resistance = Min (Mode 1, Mode 2, Mode 3) = 274.99 kN**

**Row 2 and 3 combined – column web in transverse tension**

$F_{t,wc,Rd} = \omega b_{eff,t,wc} t_{wc} k_y f_y/\gamma_{fi} = 1.0 \times 323 \times 12.8 \times 0.5174 \times 265/1000 = 566.87$ kN

**Therefore, row 3 bolt resistance on the column side = Min (274.99, 566.87) – 137.5 = 137.5 kN**

## Row 2 and 3 combined  – Beam side endplate in bending

**Mode 1 (4 plastic hinges)**

$F_{T,1,Rd} = (8n - 2e_w) M_{pl,fi,Rd} / (2 mn - e_w (m+n))$

$M_{pl,fi,Rd} = 0.25 k_y \Sigma l_{eff,1} t_f^2 f_y/\gamma_{fi} = 0.25 \times 0.5174 \times 379 \times 25^2 \times 265/1.0 = 8119.54$ kN·mm

$e_w = d_w / 4 = 39.55/4 = 9.9$ mm

$F_{T,1,Rd} = ((8 \times 48.3 - 2 \times 9.9) \times 8119.54)/(2 \times 38.6 \times 48.3 - 9.9 \times (38.6 + 48.3)) = 1037.71$ kN

**Mode 2 (2 plastic hinges and bolt failure)**

$F_{T,2,Rd} = (2 M_{pl,fi,Rd} + n \Sigma F_{t,b,Rd})/(m+n)$

$M_{pl,fi,Rd} = 0.25 k_y \Sigma l_{eff,2}$

$t_f^2 f_y/\gamma_{fi} = 0.25 \times 0.5174 \times 379 \times 25^2 \times 265/1.0 = 8119.54$ kN·mm

$F_{t,b,Rd} = (k_2 k_{y,b} f_{ub} A_s)/\gamma_{fi} = 0.9 \times 0.2705 \times 800 \times 353/1.0 = 68.75$ kN

$F_{T,2,Rd} = (2 \times 8119.54 + 48.3 \times (4 \times 68.75))/(38.6 + 48.3) = 339.72$ kN

| Compression zone | **Mode 3 (bolt failure)**<br>$F_{T,3,Rd} = \Sigma F_{t,b,Rd} = 4 \times (0.9 \times 0.2705 \times 800 \times 353/1.0) = 274.99$ kN<br><br>**Resistance = Min (Mode 1, Mode 2, Mode 3) = 274.99 kN**<br>**Therefore, the resistance of the bolt row 3 on the column side is = 274.99 – 137.5 = 137.5 kN**<br>Column web in transverse compression = $F_{c,wc,Rd} = \omega\, k_{wc}\, b_{eff,c,wc}$<br>$t_{wc}\, k_y\, f_y/\gamma_{fi} \le \omega\, k_{wc}\, \rho\, b_{eff,c,wc}\, t_{wc}\, k_y\, f_y/\gamma_{fi}$<br>$\sqrt{k_y/k_E} = 1.25; \lambda = 0.685; \rho = 1.0; b_{eff,c,wc} = 248\,\text{mm}; k_{wc} = 1.0$<br>$F_{c,wc,Rd} = 1.0 \times 1.0 \times 248 \times 12.8 \times 0.5174 \times 265/1.0 = 435.25$ kN<br><br>**Beam flange and web in compression**<br>$F_{c,fb,Rd} = M_{c,Rd,fi}/(h - t_{fb}) = 649 \times 0.5174/(533.1 - 15.6) = 648.87$ kN | Compression zone |

Summary of tensile resistances of the three rows of bolts:

Row 1 = 137.5 kN

Row 2 = 137.5 kN

Row 3 = 115.3 kN

Total tension = 390 kN<435.25 kN (compression resistance), so no reduction in bolt load is required.

Moment resistance of the connection (see Figure 5.7 for lever arms) = $137.5 \times 0.565 + 137.5 \times 0.465 + 115.3 \times 0.375 = 184.86$ kN·m>150.57 kN (assumed moment at the connection), so ok.

## REFERENCE

CEN 2005. EN 1993-1-8 Eurocode 3: Design of steel structures–Part 1–8: Design of joints. *British Standard*.

# Index